T4-AIL-872

SCIENCE FOR SANE SOCIETIES

SCIENCE FOR SANE SOCIETIES

by

Paulos Mar Gregorios

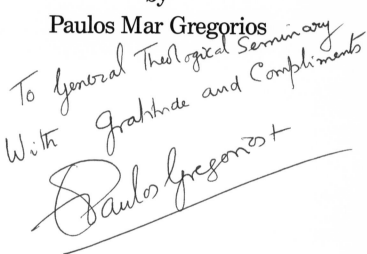

To General Theological Seminary
With Gratitude and Compliments

Paulos Gregorios +

PARAGON HOUSE
NEW YORK, NEW YORK

Published by Paragon House Publishers
2 Hammarskjold Plaza
New York, New York 10017

Revised Edition Copyright ©1987 by Paulos Mar Gregorios

Original Edition Copyright ©1980 by Paulos Mar Gregorios. Published by the Christian Literature Society, Park Town, Madras, India.

All rights reserved. No part of this book may be reproduced or transmitted in any form or by any means, electronic or mechanical, including photocopying, recording, or by any information storage and retrieval system without permission in writing from the publisher.

Library of Congress Cataloging-in-Publication Data

Paulos Gregorios, 1922–
  Science for sane societies.

   1. Science—Philosophy.   2. Science—Social aspects.
   3. Religion and science—1946–
I. Title.
Q175.P3435   1987   303.4'83   86-25282

ISBN 0-913757-70-5 (pbk.)

Q
175
.P3435
1987

Contents

SCIENCE FOR
SANE SOCIETIES

CHAPTER ONE

On the Eating Habits of Science and Faith

The World Conference on Faith, Science and the Future was not a "professional" conference. We were not there to discuss academic questions with our fellow academicians, but to engage in a common man's critique of science and technology. Professor Rubem Alves of Brazil set the tone for the Conference with his impassioned parable "On the Eating Habits of Science." Seeing scientists-technologists as wolves and ordinary people as lambs (parables are not to be taken too literally), Professor Alves made the following points:

1.   If you want to learn about wolves, do not ask them to say what they are...Most of the explanations that science proposes about itself are not only untrue, they are dangerous...
2.   Lambs know more about wolves than wolves do. A wolf is, to a lamb, what the wolf *does* to the lamb and not what the wolf thinks he is doing...These accounts, most of the time, hide its eating habits. These habits, say the philosophers, do not belong to the *essence* of science.

What if some of the voices of the "Two-third World"

were strident, irrational, emotion-packed? Rational persuasiveness was never the distinguishing mark of the prophetic voice.

> We denounce science and technology: protected by the ideology of objectivity and value-free pursuit of truth, they have been at the service of military and economic interests which have brought about great sufferings to the peoples of the Third World.
> We denounce science and technology: protected by the ideology of objectivity and value-free pursuit of truth, they have developed the most sinister instruments of death and total annihilation. We call for a halt in this situation...

Thus spoke a group of representatives from Africa, Asia, Latin America and the Pacific, a statement approved by the Conference with some amendments.

Behind the rhetoric of Two-third World participants one could listen to a fundamental criticism of scientists and technologists which could be summarized along the following lines:

(a)   Science's claims to objectivity and value-free pursuit of knowledge could be interpreted as an alibi offered by scientists to free themselves from their sense of guilt about the damage done to people by science-technology.

(b)   Scientists and technologists are guilty of having lent their services to war-establishments and quick-profit-oriented, exploiting transnational corporations and other business enterprises.

(c)   Scientists and technologists in general have not developed any ethical commitment to the welfare of humanity or the emancipation of the oppressed and exploited. Christians in science and technology have pursued success and glory and only for themselves, unlike the Son of Man who lived and died to serve the poor.

(d) Many scientists and technologists are quasi-illiterate and unreflective when it comes to the economic and political implications of their work, and even about the nature of science and technology itself.

The Two-third World prophets were accused of inconsistency in not distinguishing between science/technology* in itself and its utilization by society. The point was, however, that there could be no such distinction; that science and society are inextricably related to each other; that science and technology were born and grew up in a particular culture and a particular socio-economic structure, to whose desires and aspirations it caters and by which it is sponsored and supported.

This was the point many scientists at the Conference were unwilling to concede. An exception was Professor Jerome Ravetz of the University of Leeds, whose biting criticism of science, "both research and application, in the total, unified system on which our daily lives depend" enraged many scientists. For Ravetz, "the image of the 'scientist' as dedicated lone researcher, analogous to a saintly hermit, is now dangerously obsolete". Ravetz was indeed harsh:

> The idea of a scientist being a deceiver, or corrupt, is very nearly a contradiction in terms. How can a searcher after knowledge be a party to its distortion or suppression? If you are thus bewildered, take it as a warning that your concepts are now obsolete.

Ravetz spoke of three weaknesses in the scientific enterprise today: "First, ignorance in scientific research; second, incompetence in science-based technology; and finally, corruption in science policy."

The scientist was inclined to concentrate all the blame on science policy alone. For example, Professor Hanbury Brown, to whose thoughtful paper on "The Nature of Sci-

*The phrase science/technology is used to denote the complex system in which pure science and applied science interpenetrate and operate as a single phenomenon in present society.

ence", Professor Alves' parable was a response, took account of

> . . . the fact that in the past few decades science has been industrialized and has allied itself with power. In changing the world, it has changed itself, so that the manifest, dominant activity of science is no longer the disinterested pursuit of knowledge but the pursuit of knowledge for industry and other social purposes, such as defense, agriculture, health and so on. I won't weary you with statistics, but less than five per cent of the world's expenditure on science is now devoted to fundamental science. The vast majority of scientists are busy applying science to reach material and social goals and their work is largely controlled by governmental agencies serving national, military and civil interests and by large industrial firms serving the market.

The "Two-third World" argument was that this is not a mere incidental aspect of the scientific enterprise, and that the character and orientation of science are determined by the nature of the political economy within which science develops and operates.

The question was not adequately raised in the Conference whether these charges could be made also against the faith enterprise—especially the Christian faith. The character and orientation of the life and practice of the Christian Church have also been determined to a large extent by the political economies in which the churches flourished. Church establishments have all too often been guilty of lending support to the socio-economic establishment in which Christians lived. Fundamental and radical protest against the political economy has been seen in but a few creative pockets within the Christian community and only at certain periods of history. Churches have a longer and a more shameful record than science, of supporting slavery and bondage, oppression and exploitation, war and greed. Our theologies of work, of property, of military service, of peace and war, of the role of women, of race, and of the state have too often drawn more from the values of the establishment than from the insights of the Gospel.

Christians cannot level an accusing finger at the scientific enterprise without repenting and beginning to put their own house in order. This point was strongly made from the feminist perspective by Rosemary Reuther and Karen Labacqx. Reuther's criticism was primarily leveled at the scientific enterprise, but could be applied as well to the practices of the Church. In fact, she began with a feminist critique of Christian theology, which, according to her, "was fashioned by a fusion of these two traditions: Hebrew patriarchalism and Greek dualism". In "the religious tradition of what comes to be dominant Christianity, women can be imaged in two ways: (a) as the feminine; that is to say, as submissive, docile, receptive and sublimated (unsexual) body totally at the disposal of divine male demands, or (b) as the female: that is to say, as carnal, revolting, demonic body, which is the antithesis of the male quest for redemption through denial of his roots in the mother (mater), in matter, infinitude and in mortality".

Professor Reuther suggested that science follows the same basic pattern.

We see that a new god has been put in place of the old one, or, to put it another way, a new clerical caste in service of the political powers is replacing the old one. This new priestly class is the scientists and the technologists (see Comte and St. Simon in the early nineteenth century) and their god is the god of scientific reason. Like the old god, the god of scientific reason situates itself outside of matter, independent of it, sovereign over it (or her), knowing, dominating her from outside.

Professor Reuther further charged that both Western theology's notion of a god outside nature dominating and controlling it in accordance with his own sovereign will, and modern science-technology's fundamental attitude of standing outside nature seeking to know it, dominate it and use it for one's own sovereignly chosen ends, are reflections of the attitudes of the dominant middle class male in Western society.

It is not the case, according to her, that the Judaeo-Christian tradition proposed *dominum terrae* as the basic attitude of man to nature, and that then this notion allowed Western man to develop science and technology in order to realize that dominion over the earth. Both the so-called Judaeo-Christian notion of *dominum terrae* as well as the Judaeo-Christian sovereign dominant god are themselves creations of a white male elite ruling class in the West, insensitive to the needs and aspirations of the rest of humanity and the rest of nature. Science-technology simply inherits and puts to work this notion in a secular context.

"Historically speaking, this kind of scientific consciousness," she said, "has been the tool of a white western ruling class male elite, which has used its knowledge through technology to exploit the material resources and labor of the rest of the world (human and non-human) for the power and profit of the colonizers. This is the *key* [italics hers, P.G.] to the rapacious use of technology. The rest of the world has been dealt with as resources (material resources and labor) for the profit of the few, not as fellow beings who are to share equally in the development and benefits of the new power."

She was not arguing for "the romantic anti-scientific and anti-technological primitivism" of "back-to-nature" environmentalists. Nor would she concede that more of "conservationism" and "responsible stewardship of nature" can do the trick, for those are merely "the educated western male responses to his own self-alienation" and would amount to "the freezing of the present system of injustice".

On the contrary, Professor Reuther advocated a "thorough-going conversion of the world system" in which women should take active and equal part "in solidarity with all those who belong to the world of exploited labor". This conversion is to the new humanity in Christ, God incarnate in the flesh, the Logos or Mind incarnate in matter. This means God not outside the visible world, but as the divine matrix of all existence, real or potential, "the inexhaustible font of potential being, through which we come to be, and are continually renewed", and reason or mind "as the thinking dimension of all being...which should bind us together"—male and female, human, animal, plant, earth, air and water.

Rubem Alves' point was at least understood by the Conference, though resented and contested. Rosemary Reuther's point was not even understood by most. Many simply rejected her comments, in typical male chauvinist fashion, as a mere "women's lib" protest, irrelevant to the matter at hand. Even the second of the ten sections of the Conference, the section which had a specific mandate to deal with the interrelationship God-world-humanity, failed to come to terms adequately with her comments.

The discussion in Section II saw history differently. The report of that section (C.4) sees the early history of humanity as one in which nature (e.g., flood, fire and brimstone) was threatening to overpower and destroy humanity. Now science and technology reversed the power relations. Humanity has secularized nature, and has power over it in such a way that humanity "is now able to destroy its own species and perhaps even all life on the earth".

The Section commended "the planners of the Conference for bringing together natural and social scientists with theologians" and wanted the World Council of Churches to "promote the continuation of these discussions". It also wanted to involve our neighbors of other faiths in these discussions. But it failed to come to grips adequately with the conceptual problem which underlies the crisis of our civilization. It took a characteristically Western track, that of ethics, rather than that of deep conceptual reflection.

This tendency itself seems to be part of the sickness rather than its possible cure. We live in a secular civilization which is incapable of more than pragmatic ethical reflection. It avoids conceptual reflection for fear that this might lead to arid and unprofitable speculation. Our secular civilization is capable of coming to terms with Rubem Alves' point about the socioeconomic context of the problem of science-technology; but it still remains unwilling to heed Rosemary Reuther's point.

I believe the task ahead is that of finding a conceptual and programmatic framework in which both Alves and Reuther are given serious attention. Social scientists, natural scientists, philosophers of science, of religion, and of Marxist and other ideological persuasions, need to sit down together for some days, keeping in mind both the cry of the hungry and the cry of womankind and other marginalized people.

Such a joint effort must also take into serious account the eating habits of wolves—both of the science-technology breed and of the faith-religion breed.

Science-technology as we know it today is more than something neutral which can be *used* for good or evil. It is the fruit of a certain sociopolitical economy with its particular aspirations and specific conceptual framework of perception. It is not sufficient merely to explore the ethical aspects of science-technology as it is practiced in our world today. We need to go back to the conceptual framework, the basic perception in which the aspirations of a society are rooted. Perhaps the largest single failure of the Conference was the failure to examine those structures adequately. We shall deal with this conceptual framework after a preliminary examination of some of the ethical issues.

CHAPTER TWO

Energy

For Whom? What Kind?
At What Cost? What For?

Future energy needs, the problems of nuclear energy for peaceful use, and the need for nuclear disarmament are three separate but interrelated issues that must be examined.

ENERGY NEEDS OF THE FUTURE

How do we compute our energy needs of the future? Usually experts do it by extrapolating from the trends of the past into the future. We presume that industrial growth will continue at the present rate.

The International Development Strategy (IDS) of the U.N. Second Development Decade (1971-80) set a target of 6% per annum overall growth in the gross domestic product of developing countries and about 4.5% per annum for the industrially developed countries.

Assuming this trend, by 2025 it is estimated that the

nonfossil energy need will be five times that much—50 × 10^{16} kilojoules*—because fossil fuels by then will be able to supply only half what they are producing now.

What alternatives are available? Today 90% of world energy needs are met from fossil fuels. In the U.S.A. only 4% of the energy needs are met from hydroelectric power plants and about 12% from nuclear power plants; coal supplies 15%, gas about 30%, oil 39%.

In India, 40% of the electricity comes from hydro plants, 56% from fossil fuels, about 3.2% from nuclear power plants.

The U.S.A. is the giant consumer of fossil fuels, despite its operation of some 70 nuclear reactors. And the fossil fuel supply is finite. World gas and oil resources are being too rapidly depleted for comfort. We have no precise estimate of how much oil there is in the earth available to us. Professor David Rose gave an estimate of 2000 billion barrels. If we use up 60 million barrels a day, the total supply, including all oil still to be discovered, will last us 90 years (at 22,000 million barrels a year).

But we are actually increasing our daily consumption of oil at the rate of about 6% per year. Presently available resources of oil constitute only one-third of the potential total need, i.e., 640 billion barrels. If the present trend of oil use continues, we will have finished all the presently known stock by about 1995 A.D. The whole potential stock of the earth, if discovered and exploited, will be gone in 30 years!

OPEC countries do us a service whenever they raise prices—whatever their motive. We must drastically reduce oil consumption; there is no alternative. The situation of natural gas is similar. Only about 2500 trillion cubic feet of gas are known to exist. That is the equivalent of 420 billion gallons of oil.[1] Oil and gas are rare commodities. Their price must continue to increase, and we must look for alternatives.

*The basic unit for energy measurement is a joule or one watt-second; a kilojoule would be 1000 watt-seconds; a watt-hour would be 3.6 kilojoules or 3600 joules.

Although coal exists in greater abundance, its mining destroys the environment and increases the prevalence of cancer. Fossil fuels in general increase pollution and raise carbon dioxide levels in the atmosphere, which can seriously disrupt the biosphere of the earth in the long run. What are the major nonfossil sources? Generally speaking they are: (a) nuclear fission or fusion, (b) solar,[2] (c) hydrogen, (d) wind and wave, (e) bio-gas, (f) tidal and geothermal.

The bias in the industrialized countries, where most of the fundamental research and development goes on, has been nuclear rather than solar. Post-world-war research has concentrated on the nuclear fission technology developed during the Second World War; it had no great interest in solar. If the fundamental R & D had paid more attention to solar in the 'fifties and 'sixties, we would not be in our present predicament.

We are in a post-oil era already. Whether nuclear or solar will dominate in this era is a question largely determined by the research of twenty years ago. Unless governments are prepared to put enormous amounts of money into solar energy technology, we are almost bound to move into a nuclear energy era.

Nuclear energy

1.   Types of Nuclear Technologies

First, we should distinguish between various technologies which produce nuclear energy.

(a)   *Fission and Fusion:* Fission technology is what we have developed, following the development of the atomic bomb in the 'forties. In fission, energy is released by splitting a heavy nucleus of an element (e.g., uranium) into two smaller nucleii of other elements.

Fusion is the process in reverse—the formation of a heavier nucleus from the fusion of two lighter ones (e.g., deuterium and tritium). Here technology has lagged behind. We still have not got around to creating the necessary high temperature under controlled conditions to produce fusion in such a way that the energy can be constructively used. We have done much better in the use of fusion in nuclear weapons and warheads like the hydrogen bomb. Fusion technology is reputed to be safer and cleaner; i.e., it has fewer hazards of radioactive wastes, fuel storage, fuel hijacking, etc. But it is still a long way off. According to the experts the entire energy needs of the U.S.A. can be met by fusing 10 kg of deuterium per hour. They expect that the first experimental fusion reactor will be operative by 2005 A.D., and the first commercial reactor by 2025 A.D. There are other technical problems to be resolved, beyond the production under controlled conditions of a hundred million degree hot plasma—new materials to withstand constant high energy neutron bombardment, maintenance and repairs, as well as the handling and recovery of tritium. These problems are soluble. But then yet unseen problems may very well arise.

(b) *Burners and Breeders:* A burner reactor uses the fuel only once, producing some fissionable material, but consumes more fuel than it produces. A breeder reactor produces more fissionable fuel than it consumes and the new fuel can be used in a continuous process. Since the uranium supply is limited, breeder reactors are preferred by most nations. A *Light-Water Reactor* (LWR), which uses ordinary water as a coolant to moderate the heat generator, is distinguished from a *Gas-Cooled Reactor* (GCR), which uses gas (often generated from graphite) as the coolant; it is also distinguished from a *Heavy Water Reactor* (HWR), which, by using water containing more deuterium (the heavier part of hydrogen) as the coolant, slows down neutrons and permits use of unenriched natural uranium.

The *Liquid Metal Fast Breeder Reactor* (LMFBR) can use either the uranium cycle or the thorium cycle. Uranium 238 and thorium 232 are rather abundant on our earth. The U.S.A. has at least 3.7 million tons of known natural uranium deposits, while Western Europe, having only less than half a million tons, is at a disadvantage. Australia has one-fifth of the total known world reserve. Other deposits are in the Soviet Union, in Namibia, South Africa and Greenland. Uranium technology is more developed than the technology of the thorium cycle, partly because thorium is more abundant in developing countries, while uranium deposits seem to favor developed countries and their colonies like Namibia and South Africa. Thorium cycle technology serves the interests of the Two-third World, but they have less money to invest in research.

Natural uranium (U^{238}) is fairly stable. It yields about 1% of the fissionable isotope U^{235}. An *enrichment plant* will raise that yield to about 3% to 4% for reactor fuel, but can use almost 90% of U^{238} for making bombs. A *reprocessing plant* makes plutonium, the man-made fissionable element. A few thousandths of a gram of plutonium inhaled causes fibrosis of the lungs leading to death, and plutonium can continue to be radioactive for 24,000 years (some say half a million years). And the cost is high: Brazil, for example, has a contract with West Germany to build eight reactors, one enrichment plant and a reprocessing plant at a total cost of five billion dollars.

Forty-four countries of the world's 150 nations are now committed to developing nuclear energy. Supplying nuclear technology is big business, and transnational corporations are heavily involved in them. Both Taiwan and South Korea now have a nuclear reactor, and many more developing countries are buying.

2.   Some Technical and Ethical Issues in Nuclear Technology

(1) *Pollution.* Some of the nuclear power facilities (reactors and reprocessing plants) emit even normally very high levels of carbon-14, tritium (H_3), krypton-85, iodine-129, and perhaps also cesium-137 and strontium-90. These elements have "half-lives" or high radioactive periods of ten years to several million years. Some of them accumulate in the food chain and cause serious damage

to people. Plutonium 239 has a half-life of 24,400 years and will still be emitting alpha radiation after 250,000 years.

(2) *Waste Management.* Used reactor fuel is highly radioactive and goes on being active for thousands or hundreds of thousands of years. Wherever we store it, it is difficult to make sure that it will not pose major radiation hazards for present or future generations, but safer technology is now emerging.[3] A typical large reactor produces 30 to 40 tons of spent fuel a year. All reactors produce some plutonium.

(3) *Fuel Transportation.* Nuclear fuel is expensive and dangerous. Accidents can occur during transportation, posing threats of radiation for unsuspecting people. Nuclear fuel can also be hijacked and used for subversive activities.

(4) *Plant Accidents.* The Three Mile Island (near Harrisburg, Pennsylvania) accident put everyone on guard. This incident could have caused a catastrophic melt-down, imperiling life on the whole planet. Largely because of human error, significant quantities of the deadly poisons lesium-137 and strontium-90 were released into the air and water around the plant. People are understandably averse to such risks of exposure to radiation and poison. There have been several such accidents already.

(5) *Proliferation of Nuclear Weapons.* The acquisition of technology for developing nuclear energy for peaceful purposes also gives access to nuclear technology for warfare. Although more than 40 nations are committed to developing nuclear reactors for peaceful use, it will be difficult to keep them from developing nuclear weapons also.

(6) *Security Requirements.* Since nuclear fuel and technology are easy to misuse, especially in the hands of antisocial or revolutionary groups, nuclear plants require extra security measures and greater surveillance of people working with nuclear fuel and technology. Such action augments the "police state" character of our modern societies.

(7) *The Future of Our Industrial Civilization.* The central ethical issue of nuclear energy for peaceful use lies

in the domain of the kind of future civilization that we should choose to have. As Professor Jean Rossel of Switzerland has said:

> Whether we like it or not, nuclear energy and its industrial organization, along with space exploration and its excesses, have become a sort of extreme expression of our simultaneously arrogant and fragile technological society. Rightly or wrongly, the nuclear industry now represents, in the eyes of the ordinary man who is still sensitive to traditional values, the quintessence of risks and dangers, both short and long term, difficult to assess, and therefore all the more distressing.

In other words, the exploitation of nuclear energy is another symbol of the trend towards gigantism and centralization in the structure of our economies. Technology as a source of unlimited power for man should not grow too far ahead of human capacity to keep that power under control. But in our industrial civilization, technology seems to have acquired a momentum which seems to be already out of human control. Many young activivists in the anti-nuclear campaign therefore argue that their opposition to nuclear energy is in fact an invitation to call a halt to this uncontrolled development of technology in order to have a little breathing space to consider and develop alternate styles of living a civilized human life. These young people do not always simply advocate a return to precivilized existence. They are arguing for a society in which there is less consumption of manufactured commodities and more consideration for the environment which is necessary for the survival of life on our planet; for the development of new forms of science and technology to promote a better quality of human life and society, and for finding ways of living together with each other and with the rest of creation which would foster rather than hinder the growth of human dignity and freedom, justice and peace, love and joy.

This view sometimes uses dubious arguments, mostly in order to attract attention. For example, Professor Rossel invoked a paper read in 1958 in a Radiation Symposium in Switzerland, which argued that the disappearance of the giant Saurians of the Mesozoic era was caused by a slight rise in natural radioactivity on our planet which became fatal to the delicate biological equilibrium of the bulky bodies of these prehistoric giant creatures. The associated thought is that large-scale exploitation of nuclear energy and the consequent increase in radiation may cause a similar imbalance in the human biosystem which could spell the end of the human species. Such arguments may not carry conviction with scientists, but can invoke highly irrational anxieties and fears in the minds of ordinary persons.

On the other hand, the protagonists of nuclear energy also use dubious arguments. For example, they argue that nonnuclear energy sources are not commercially viable. All this probably means is that while the large-scale development of scientific technological research for nuclear energy for peaceful use is already more than 40 years old (after Hiroshima-Nagasaki, after the Second World War), an equally large-scale research effort for solar energy development has not yet begun. And it usually takes 20 to 30 years before such research produces something commercially viable.

The other argument uses comparative statistics on deaths caused by nuclear energy, and by other forms of technological activity like the mining and burning of coal and other fossil fuels, or by the driving of automobiles. Professor Rose provided statistics from the U.S.A. to show that out of 8270 cancer deaths in the U.S.A. in 1977, 3960 were caused by ambient exposures like cosmic radiation, 2960 by medical and dental X-rays, 660 by radiopharmaceuticals, 600 by technologically enhanced natural radiation from fossil fuel powered plants and inactive uranium mines, 80 from fallout, and only 9 from the uranium fuel cycle. Nine out of 8270 is just about 1/10 of one percent. Clearly we should be more worried about dental and medical X-rays and fossil fuel burning power plants than about nuclear reactors.

Dr. Rose would argue that the burning of fossil fuels may have more catastrophic consequences for the biosphere than the development of nuclear reactors. There must be a certain truth in the argument that carbon-dioxide levels in the atmosphere are greatly on the increase due to large scale burning of hydrocarbons in our industrial civilization. This may cause, through the famous "greenhouse-effect" (carbon-dioxide and water-vapor in the atmosphere absorb the heat radiated from the earth, and provide a warm envelope to keep the heat close to the earth's surface), serious climate dislocations which can have ruinous consequences.

But this cannot be an argument for nuclear energy. It is an argument for radically slashing down our fossil fuel consumption, even if the industrial system has to be fundamentally reconstructed at great cost, in order to make this slashing down possible.

We have as a civilization become quite used to the fact that at least 100,000 people die each year in our world from automobile accidents. Yet we do not give up the automobile. Even with a thousand nuclear reactors in operation, the annual death rate from nuclear radiator-related accidents is unlikely to approach the automobile-related accident death rate. It is, then, not the possible death rate that is the heart of the nuclear energy debate.

When I decide to drive an automobile, I know I am exposing myself, my fellow-passengers, and others on the road or in other cars to the risk of an accident. There are some differences, however, between the risk of driving an automobile and that of building a nuclear reactor. What are these differences? We might mention three:

(a) *The magnitude of the risk:* I am better able to imagine the possible extent of damage I can cause to myself or to others in my driving a car. The risk of a reactor accident or fuel disposal hazard is more difficult to assess beforehand. The failure of one human being or one action of nuclear waste disposal is much more difficult to imagine or assess beforehand. People feel much more vulnerable or defenseless against a nuclear accident in plant or waste disposal. There is a general feeling, whether justified or not, that my freedom to exist is much more radically threatened by a nuclear reactor than by a host of automobiles, or by coal mining.

(b) *The range of the risk:* The hazards of nuclear energy constitute a threat, not only to me or to the present generation, but also to future generations. But one could also say that the carbon dioxide increase phenomenon caused by excessive burning of fossil fuels can be just as threatening to future generations. Here the objections to nuclear energy could apply equally to the burning of fossil fuels.

(c) *The risk of the reinforcement of present patterns of industrial civilization:* This is perhaps the key argument, and perhaps the most potent one, against the use of nuclear energy. By opting for large-scale investment in nuclear energy, we are opting for a reinforcement of the present problematic pattern of industrial civlization. We will find it much more difficult to change course once we have invested a great deal in nuclear energy. But certainly not opting for nuclear energy will not automatically change the course of our industrial civlization. At the present moment we have not found effective means for changing that course. This may only mean that we must set our hearts and minds to a more resolute program to effect such change and should not render that task more difficult by large-scale investment in nuclear energy. The central issue then is not nuclear energy but the development of an alternate civilization.

ALTERNATE SOURCES

Whatever may be the decisions about nuclear energy by individual nations, no nation will choose to depend entirely on nuclear reactors for its energy needs. It will have to be a package, containing possibly fossil, solar, hydro, and others. Hence we must look at alternate sources of energy—both renewable and nonrenewable.

Fossil fuels are nonrenewable; other sources like wind,

sun and wave are renewable, i.e., in constant supply. Nuclear energy produced by breeder reactors can be placed in an intermediate category, since more fuel is produced than consumed.

In view of the limited supply of fossil fuels, everyone thinks that we must put all our efforts into the renewable sources, especially the most abundant source, solar energy. The dispute is now between nuclear and solar, but the debate needs a lot of further clarification—especially between nuclear energy and solar energy produced by profit-minded commercial corporations in a market economy, and that produced by a genuinely people-based and properly administered socialist state. Many of the problems encountered in a market economy structure are not faced or felt in the same way in a socialist economy.

Today all of us are victims of lobbies in the market economy world, parties interested in promoting either nuclear or solar energy for the sake of the corporations' profit. And we are gradually learning to take "expert estimates" of costs and consequences of either type of energy with a grain of salt. We suspect that behind each expert there may be an interest lobby; though the experts themselves may not be directly linked to such a lobby, the publicity and promotion given to their views are likely to be lobby-limited.

It is clear that solar energy is more abundant in the tropics; but it is there that energy for space and water heating is less in demand. Solar energy for space and water heating will become big business in the temperate climates, where most of the affluent societies are now located. For most of the developing countries, electricity for industrial and domestic needs is the first priority; here a great deal more money will have to be invested to bring research to the point of commercial viability for solar energy. Fundamental research will have to be undertaken in the realm of Technical Cooperation Among Developing Countries (TCDC) if this is not to become another means of exploitation of the developing countries by the developed.

THE MORATORIUM ISSUE

The Conference debate did not focus on this issue, mainly because the controversial proposal for a five-year moratorium on new nuclear plants occupied the center of the energy debate. It was somewhat disconcerting to many nuclear experts that the Conference recommended that governments should:

> Immediately introduce a moratorium on the construction of all new nuclear power plants for a period of five years. The purpose of this moratorium is to encourage and enable wide participation in a public debate on the risks, costs and benefits of nuclear energy in all countries directly concerned.

Even if the moratorium itself is not imposed by the governments, the recommendation will serve some purpose if a public debate with wide participation can be initiated in many countries.

The conference adopted other equally significant recommendations in the energy field. To mention only a few:

(1)   That, in the interests of energy conservation, a "Fuel Pledge" be internationally introduced, say, something like "I pledge myself to save fuel and electricity at home, at work and at leisure, and to help to make more available for those whose basic needs are not being met."

(2)   That we "identify ethical criteria by which the social impacts of energy technologies must be assessed and insist that in setting energy policy, such criteria be given equal weight alongside technical and economic factors."

(3)   That the "pollution of the environment by an excess of carbon dioxide, radioactivity and other products of the extraction and combustion of fuel be substantially researched and kept to the minimum that is technically feasible."

The moratorium issue was clearly one of the most controversial in the Conference. Two questions can be briefly considered here:

(1)   What purpose does such a moratorium demand/serve?

(2)   Is the position taken by the Conference fundamentally different from the position taken by the Central Committee of the World Council of Churches?

1.   *Purpose of Moratorium Demand.* It is unlikely that governments will respond to such a demand from the WCC, despite the wide publicity generated by the Conference debate. The demand, after all, is only for a limited time (five years) moratorium. Its purpose is clearly stated as that of promoting a real public debate with wider participation of experts and nonexperts. It is likely that more people will take an interest in the on-going debate as a result of the Conference debate, whichever side they may have voted for in the Conference.

2.   *The WCC Position.* Two basic elements in the WCC position taken by the Central Committee (Jamaica, January 1979) remain unaltered: (a) that nuclear energy is a conditional good, i.e., that it can serve a purpose beneficial to humanity provided the necessary safeguards are developed and enforced; and (b) that, for the present at least, all energy options are to be kept open and therefore that nuclear energy cannot be unconditionally rejected as evil in itself. The future debate should focus on both the questions, i.e., adequate safeguards and what the development of nuclear energy does to the shaping of future society.

CHAPTER THREE

Bio-Ethics[1]

*Genetic Manipulation—Social Biology—Bacterial
Research—Social Control of Science and Technology*

BIO-ETHICS IN GENERAL

The Conference did not attempt to deal with the whole
spectrum of issues in medical and biological ethics. It
focused instead on "Theological and Ethical Issues in
the Biological Manipulation of Life." More specifically it
dealt with:

(a) genetic manipulation or engineering;

(b) behavior control;

(c) prolongation of life of terminally ill patients;

(d) psychological manipulation through media, ad-
vertising, etc.;

All these issues raise important theological questions
for which no clear and undisputed answers are readily

available. The Conference section concerned with these questions took a pragmatic rather than conceptual or theological approach to these questions. There was general agreement that the industrially developed nations faced these questions with greater urgency than the developing countries, where the major point of interest was the more just and equitable distribution of scarce medical resources. My experience in India shows that in developing countries, while the medical profession shows some interest in these questions, the educated general public is hardly aware of them. The reasons for this lack of interest should be investigated.

The section approached these questions from the perspective of criteria for decision-making, but ultimately they had to recognize that the criteria were integrally related to certain theological-anthropological commitments of which people are insufficiently aware.

The subtitles of books on such issues are often revealing. To take just two examples, Mary Shelley's work on Dr. Frankinstein is subtitled *The New Prometheus*, while Joseph Fletcher's *The Ethics of Genetic Control* (Anchor, 1974) has the subtitle *Ending Reproductive Roulette*. The popular image of the scientist in general and the genetic researcher in particular, as a Dr. Frankinstein who plays God by trying to create monsters, is opposed to the Fletcher view of biological research as part of humanity's fulfillment of its God-given vocation to be a cocreator with God in shaping humanity in God's image.

Theologically, the questions could be put thus: (a) should we accept humanity's genetic and biological endowment as given, or should we try to "improve" it by artificial means? and (b) if we accept the second alternative, what norms and criteria should be set up to control the orientation of this "improvement"?

The Conference generally took the view that we are both creatures of God and cocreators with him, and therefore have a responsibility, within the limits of the possible and the desirable, to "improve" the biological and genetic endowment. But the Conference was also careful to point out that such "improvement" should not overlook the necessary sensitivity to the life of each human person.

Some other theological issues that underlie the debate are:

(a)   Is alleviation of suffering the highest criterion for decision-making?

(b)   What is meant by a "defective" human being, or genetic "defects" in an embryo?

(c)   Do we take normal decisions about prenatal abortion of "defective" embryos in terms of a cost-benefit analysis?

(d)   Do embryos have rights—e.g., the right to life? Are these rights adjudicable or merely moral? When they come in conflict with the "rights" of parents, how do we make decisions?

(e)   Do mentally retarded or otherwise genetically "defective" people have a right to progeny?

The idea of "defect" in a human being is hard to determine. One could say that blindness is a defect, but we do not deny the blind man's right to live or his full dignity. A mentally retarded person is still a human person, and only Hitlers advocate the extermination of all mentally retarded persons.

The alleviation of suffering is a good thing. But is it in accordance with Christian teaching to say that a mother can opt for the annihilation of an embryo that is likely to cause her pain and suffering? Does not the biblical understanding of suffering go deeper than the question of alleviating it?

The Conference did not deal with all these questions. On some issues it gave clear and unambiguous answers. On others it merely left the issue to be discussed by the churches.

1. *Artificial insemination.* The section report clearly stated: "the practice of artificial insemination with husband's sperm . . . is morally unobjectionable." It also stated the fact that many Christians object to artificial insemina-

tion by a *donor's* sperm on the ground that it is a violation of the marriage bond; but the section document holds
that "others do not now share this position." The section
report goes on to point out the need to regulate by law the
growing institution of semen banks which have to maintain certain standards. The section did not, however, question the whole institution of frozen semen banks. It was
concerned about the legal protection of the social standing
and inheritance rights of children produced by Artificial
Insemination Donors. It took a clear stand against AID
for unmarried women.

2. *Abortion.* The section document was predictably
noncommittal on the question of prenatal abortion of embryos. It recommended the setting up of a commission to
go into the issues like:

(a)   Is there any moral difference between abortion
and infanticide?

(b)   If abortion of genetically defective embryos (detected by amniocentesis) is more widely accepted,
what would be the long-term effect on the moral values
of society?

(c)   How do we pastorally help Christian parents who
do believe that all human life is a gift of God and yet
opt for abortion in their own case?

3. *In vitro fertilization of human embryos.* The conference document did not question the moral rightness of
in vitro (in a glass) fertilization of human embryos formed
of parents who cannot otherwise have children. It did
question it on the ground that *in vitro* fertilization is enormously expensive and therefore violates the principle of
equitable distribution of scarce medical resources and
skills.

4. *Cloning.* The gap between scientific knowledge
and technical feasibility has become very narrow in the
case of cloning of human beings and other animals. Exact
replicas of a given individual or of a fertilized ovum can
be made, though it is rather expensive to do so. Again, the

moral issues raised by possibilities of cloning were not analyzed in detail. The crucial issue of monosexual reproduction by cloning and its ethical justification does not appear in the documents.

GENETIC ENGINEERING

This is, of course, the dramatic new possibility. Things are moving so fast that ethical reflection is hardly able to catch up with the new possibilities. Only twenty-six years ago Watson and Crick analyzed the chemical structure of the compounds of DNA, the master molecule in most genes. In 1975, Nobel Laureate Har Gobind Khorana created a biologically active synthetic gene. On November 7, 1977 a team of California scientists "created" five milligrams of somatostatin, an important human brain hormone, through combining three synthetic genomes and thus creating a new artificial gene.

It is no longer idle to talk about humanity's taking over from nature the job of business manager of the process of biological evolution. Recombinant DNA technology now enables humanity to radically alter the genetic endowment of a person, as well as possibly to create new "species" not found in "nature".

And the pace quickens. Professor Jonathan King of MIT told us that the 1978 budget for biomedical research in the U.S.A. is about 3 billion dollars—1000 times the federal expenditure on biomedical research in 1948. The achievement so far is phenomenal and impressively fast.

We now know and understand:

(a)   the chemical structure of the key genetic component DNA;

(b)   the organization of the genetic material in linear segments, or the genes;

(c)   that genes constitute blueprints for the structure of protein molecules, the key components of living cells;

(d)   the role played by the thin membranes which divide the cells into compartments;

(e)   the organization and functions of the complex ribosomes or factories for assembling new proteins according to the instruction of the genes;

(f)   the technique for incorporating segments of DNA derived from one organism into the cells of another organism;

(g)   the technique for cloning these synthetic genes;

(h)   the technology for creating new strains of plants and microorganisms;

(i)   the technology to correct inherited blood diseases like sickle cell anemia by removing bone marrow cells and replacing them with healthier cells.

The above is only a partial list. What ethical questions do these possibilities raise?

It is clearly recognized that they do offer man very effective tools for preventing and curing disease, increasing agricultural production, for generating new energy (bio-mass), and so on. But the negative possibilities are somewhat frightening, for example:

(a)   inadvertent or intentional creation of pathogenic bacteria strains; their possible escape or release from the laboratory;

(b)   the increased possibilities of biological warfare, climatological or environment-disruptive warfare, use of pathogenic bacteria for blackmail, hijacking, etc;

(c)   genetic engineering on criminals, prisoners, revolutionaries, etc., which may be ethically unacceptable even if the prior consent of the person is obtained.

Even more problematic is the fact that in market economy systems like the U.S.A., corporations are moving into this field with all deliberate speed. Professor King mentioned International Nickel, Standard Oil, Imperial

Chemical Industries, and the Eli-Lilly Corporation. They are investing substantial sums to exploit the new technologies for commercial purposes. Some of these are clearly beneficial—e.g., the creation of more productive strains of cereals like rice or wheat, and developing strains of blue-green algae which fix nitrogen for fertilizing rice paddies.

But companies are now moving in to establish patents on some of these technologies and, by claiming royalties, seek to exploit the populations of developing countries. A company that funded the scientific research that resulted in a particular technology may claim that the knowledge yielded by the research is its private property. Should not all technical knowledge be the property of humanity rather than of particular individuals, groups or corporations? This is a major ethical issue.

Even more complicated is the question: What principles should regulate the orientation of research itself? Can we say that all knowledge is valuable in itself and that no restrictions at all should be placed on scientific research? Should the scientist be allowed to do research which might cause damage to society—e.g., the development of new bacterial strains?

Are there some things that we value about man and nature that set limits to what is normally permitted in scientific investigation? What are the criteria?

Dr. James Gustafson, formerly of Yale, put it this way:

> A scientist has no right to intervene in the natural processes in such a way that he might alter what men believe to be, and value as the most distinctive human characteristics ...A scientist has the right to intervene in the courses of human development in such a way that the uses of his knowledge foster growth in those distinctive qualities of life that humans value most highly, and remove those qualities that are deleterious to what is valued.[2]

One can agree with his general principle, but the problems of actual implementation are enormous. Humans do not agree on what is to be valued and on what are the most distinctive human characteristics. Sight, for example, is something human beings value very highly. Some forms of blindness may be hereditary. By sterilizing people who are congenitally blind, we may improve the sight capacity of the race as a whole. But this would be at the cost of certain other values which human beings value very highly— for example, the right to have children of one's own. The same principle applies to those with other genetic defects (e.g., mongolism) or congenital criminal tendencies (e.g., the X-Y-Y syndrome in chromosome structure).

The ethical choice is not between right and wrong, clearly defined, as Karen Lebacqx quite clearly pointed out at the MIT Conference. The choice is between two sets of values, values cherished by the same person, or by different persons, or by society and the individual, or by society and a group of individuals, etc.

We have not yet come to the point of giving clear criteria for making decisions in the light of conflicting values. Perhaps we may never get to that point. The problems of the methodology of decision-making will be dealt with in a later chapter.

SOCIAL BIOLOGY

The issues around social biology stirred up considerable debate in the MIT Conference. Social biology, associated with the names of E. O. Wilson, Konrad Lorenz, and Desmond Morris (*The Naked Ape*), seeks to understand human social behavior in terms of the genetic heritage common to all primates and even all animals.

Critics of the sociobiologist view were charged, by the advocates of that view, of misrepresentation. The moral problem, however, related to the question of human responsibility for human behavior. If certain behavior patterns like aggression or escape are genetically determined by the circumstances of our evolutionary heritage, then how can people be held accountable for what they cannot but do? Granted that human beings are naturally endowed with aggressive or fugitive tendencies, do we still not have some responsibility for controlling some of our natural tendencies in the social interest? If this were not so, no sexually attractive woman could walk safely on our streets. The interesting questions that emerge are (a) the extent to which human beings are responsible for their social behavior, (b) to what extent we can use chemical alteration of our natural endowments, (c) what criteria are available for orienting such alteration, and (d) what means are to be used. There are certain means which have been traditionally used, like fasting for example, to bring certain "natural"drives under control. Some societies have used drugs like Peyote for altering consciousness and experiencing other aspects of reality perception. Many societies approve the use of alcohol for overcoming inhibitions, for inducing temporary states of euphoria, and for promoting greater sociability. Where is the dividing line between the use of temporary stimulants and narcotics like coffee, tobacco or alcohol, and other drugs like marijuana or LSD? As the document of Section IV puts it:

> A South American revolutionary and an Indian committed to nonviolence might both share the same kind of genes for aggressiveness, but the behavioral consequences of these would be totally different. The notion, therefore, that there are genetical determinants of human personality which sociobiology might progressively reveal is seriously misleading.

Granting that conclusion, the fact remains that more detailed knowledge of our biological and evolutionary heritage may help rather than hinder the process of bringing human social behavior under more conscious control. The difficulty remains, however, that the theories of social biology remain far from scientifically demonstrated. While the tendency to explain all human social behavior through an analysis of our evolutionary heritage may be seriously misdirected, there can be little justification to forbid deeper studies of our sociobiological heritage.

SHOULD SCIENCE BE CONTROLLED?

The question of social control of science arises mainly in nonsocialist societies. In socialist societies, at least in principle, the Party, on behalf of the people, controls the development of science by laying down policy, by controlling funds, and by strict supervision of academic institutions. In Marxist thought science and technology are part of a system: science-technology-economy-Man. The whole socioeconomic process is centrally controlled and science and technology constitute important but never independent units in social production. Man himself is seen as integrally related to the system of social production, in which again science-technology is an integral part. Man created science-technology, but it feeds back to shape man.

Yet man is not simply a passive object to be shaped by science-technology or by social production. He is a conscious and free agent who can reconstitute himself by restructuring science/technology. The Research and Development sphere is where this restructuring has to be more consciously applied.

Science policy, as well as its implementation, is thus centrally controlled by the Party on behalf of the people. We know very well that this system is subject to serious abuse, precisely where the party loses touch with the people and no longer represents their best interests or fails to keep answering to the people.

Socialist countries are now resorting to a systems analysis type of social control. Dr. Lech Zacher, Head of the Section for the Scientific and Technological Revolution in the Polish Academy of Sciences, puts it this way:

> From the point of view of the necessity to control the process of the scientific and technological revolution, civilizational potential may be regarded as a system; and the spheres of Science and Technology, Economy, and the Sphere of Man's activity can be treated as its subsystems...The sphere of Science can serve as an example of a subsystem which stops being subordinate and turns into one of domination... For the needs of controlling of the processes of the scientific and technological revolution it is indispensable to define (by means of various parameters) the nature and energy of mutual impact of the subsystems on one another, as well as to specify the means and methods of conscious human actions performed within the frames of individual subsystems.[3]

The real difficulty with this cybernetic control of society is that it becomes so much the more impersonal and out of the reach of the common people to understand, monitor, or control. Besides, the programming of the network of Science-Technology-Economics-Man interrelationships into the system is not at all easy. Socialist societies have not come close to achieving such a systemic control system, though they have conceived it.

The situation is quite different in nonsocialist countries. There are government agencies in many countries which monitor food and drug manufacturers, pollution control, and various other scientific activities, but any overall control of scientific R and D seems practically impossible in the developed economies, where private interests have so much stake and say in matters of scientific research. The thinking citizen in nonsocialist countries begins to develop a measure of insecurity in the absence of any reliable agency to defend people against exploitative or harmful use of science and technology by individuals, corporations or governments. In socialist countries governments seem to be the main source of potential misuse of science and technology against the people.

We will certainly need, even in a country like India, rather strict control of Research and Development in certain sectors—primarily nuclear technology and genetic technology. But in a vast and varied land like ours the implementation of such control is bound to prove difficult, especially as corporations and individuals acquire the know-how and resources necessary for fundamental R and D in these fields. It is perhaps harmful to try to control all forms of scientific R and D. Science should be free. But people should be free also, at least from being seriously damaged by science and technology. Nothing short of vigilance can be effective in this area, combined with necessary legislation and effective implementation.

CHAPTER FOUR

Human Existence in Danger?

THE SUSTAINABILITY OF THE ENVIRONMENT

Human Impact on the Environment

The human impact on the environment can be measured mainly in terms of the following:

1. the density and rate of growth of population;

2. the rate at which finite resources are consumed;

3. the rate at which air, water, soil, etc. is polluted;

4. the degree of ecological upset caused (deforestation, carbon dioxide increase, etc.);

5. the degree to which the threat of a nuclear war is increased.

Each of these factors, as they increase, adversely affects the sustainability and the ecology of our planet. We have come to a stage where the biosphere which humanity has inherited and the technosphere which humanity has

33

created are not only out of balance, but dangerously in conflict. The word *ecology* comes from the Greek *oikologia*, which means the science of human habitation. How humanity transforms the environment by living in it and from it is the key question.

Quite often we forget the fact that the biosphere—this layer of the surface of the earth where the conditions for plant, animal and human life are available—is a fragile film which can be easily damaged. People in the developed world are often tempted to dismiss this problem, but this is a reflection of both our arrogance and our ignorance.

POLLUTION

Let us take the problem of pollution first, since it is perhaps the most easily manageable. It is one of these instances in which technology can solve the problems it has created.

Today most sensible industrial enterprises are reckoning with a cost of 2 to 4% of capital outlay for pollution abatement. That is, of course, an enormous cost. There are some kinds of pollution for which no adequate commercial abaters have yet been found—e.g., sulphur, nitrogen oxide, hydrocarbons and carbon monoxide. But most other particulate air pollutants are now manageable at some cost. We are still looking for abaters for pesticides, agricultural pollutants and radioactive waste.

The problem here is that governments are not yet sufficiently free, especially in market economy countries, from the lobbying pressure of corporations, to impose and enforce adequate pollution-abatement measures.

Our lake and river waters are badly polluted. But with determination and sufficient financial investment, these can be cleaned up and the pollution rate significantly reduced if we can find alternate means to dispose of human and industrial waste.

It is common knowledge that automobiles are a major source of pollution. But this problem can be managed, provided there is sufficient public outcry.

RESOURCES

Pollution and resources have some interconnections. For example, if the water of rivers and lakes is badly polluted, the food supply can be seriously affected by damage to aquatic life. Clean water is not just a natural resource; it is also a commodity which people consume more or less directly.

Food is perhaps the most important resource, along with air and water, for human survival. With available technology, the productivity of land in cattle and crops can be increased up to 300% by the year 2000—especially in developing countries. A 60 to 100% increase is possible even in developed countries.

Provided sufficient funds and technology are invested, at least 229 million more acres of land can be made arable —this is 30% of the arable land in 1970. The cereal potential of land is about 8 tons per acre per year. But most land in Asia is still producing about 2 to 2½ tons per acre. The food problem is not as insoluble as people think. Sometimes much is made of food scarcity in the world merely to justify an inordinate increase in food prices.

Mineral resources pose the most difficult problems. If present rates of industrial growth continue, and they have to continue for at least three quarters of the world, the following increases in mineral demands can be anticipated by 2000 A.D.:

Copper	580%
Lead	630%
Iron Ore	570%
Nickel	520%
Petroleum	620%
Gas	550%
Coal	600%

With the development of recycling technology, the problem will be abated to a very small extent. Our coal and iron ore supply for electric batteries (even solar photovoltaic cells need lead and cadmium) and other industrial uses will run out by the end of the century. Tin, tungsten, asbestos, fluorine, mercury, phosphorous— all these may be in very short supply. Even with new discoveries of deposits and new technologies for consuming less, we are likely to run into serious trouble by the end of the century.

POPULATION

Population control is the most well-known issue. Especially in the developed countries there has been much concern about the excessive growth rate of population in the developing countries. As far as resource consumption is concerned, it is the population of the developed countries which constitutes the bigger problem.

It is now generally recognized that there is an inverse correlation between population and standard of living. An increase in standard of living in the developing countries at first leads to a fall in the death rate and consequently to an increase in population. But as standards rise, the birth rate begins to fall, at least in the second stage. Although governments in developing countries should invest a great deal of money in population control, it is hoped that eventually the assurance of a dignified standard of living for the poor of the world will be a more effective factor in controlling the birth rate and, consequently, the population increase.

But until 2000 A.D. the present overall rate of growth in population is likely to continue. Only by 2000, the Leontieff Report says, some stability will be achieved in the overall world population growth rate, provided the standard of living continues to rise at an appreciably higher rate in the developing countries.

Unfortunately the situation in many developing countries remains tragic. The overall growth in the agricultural and industrial output of developing economics is more than offset by the population growth, so that there is no appreciable increase in the per capita consumption. The plight of the poor is made even worse by the fact that the overall increase is largely absorbed by the growing mid-

dle class, with the result that, in a country like India, 60%
of the people have experienced little tangible improvement
in their subhuman living conditions.

Population regulation is a major factor in assuring the
sustainability of the human habitat, but it is certainly not
the only important factor.

The section report rightly emphasizes the fact that
"social sustainability" is more important than mere regulation of numbers:

> In seeking that goal of ecological sustainability, the goal
> of social sustainability must be sought with equal fervor if
> the beneficial population levels are to be achieved within the
> framework of a just, participatory and sustainable society.
> This would include such goals as sustaining a life of dignity,
> meaning and human worth . . .

ECO-BALANCE

The section report of the Conference paid considerably
more attention to the rural-agricultural aspect of sustainability than to the industrial-urban aspects. This may have
been because they thought that the threat to eco-balance
from industrial and military technology was adequately
known. The section called for a new land ethic and quoted
from Lester Brown's World Watch Paper No. 24 (October
1978): "The times call for a new land ethic, a new reverence
for land, and for a better understanding of our dependence
on a resource that is too often taken for granted."

Since most of the world's people still live in rural areas,
this emphasis on land, forest and water is perhaps justified.
Mindless irrigation leads to erosion of land; thoughtless
deforestation leads to changes of climate and to desertifi-
cation; unplanned cropping systems lead to impoverish-
ment of land; by the end of the century, it is feared, only
2% of the earth's surface will be useful for sustaining the
needs of six billion people, since more than 70,000 square
kilometers of agricultural land are lost every year. The section report gives some alarming figures indeed:

In historical times, more than half of the earth's arable soil resources have been lost. Annually half a ton of top soil is irrevocably lost for every man, woman and child now living. Thirty percent of the remaining half of our soil deposits is predicted to be lost by the end of this century. . . during a time when human populations will increase by this same percentage.

The disruption of the atmosphere, so vital to the biosphere, by pollutants calls for vigilance on our part. As the section report said:

It is both prudent and ethically necessary to carefully monitor the effects of discharging large amounts of gaseous, solid and radioactive materials into the atmosphere. Whether the pollution is from industrial plants, power stations, automobiles, dust from agriculture, fertilizers, aircraft or other sources, it threatens the protective ozone layer, the thermal balance of the earth's climate, and affects all forms of life on earth.

The section therefore questioned the ethics of valuing industrialization more than human health.

The Antarctic region received special attention from the section. The Antarctic continent and all the islands south of 60 degrees latitude have a key place in the regulation of the world's climate and ocean current circulation.

Thirteen nations have now signed an Antarctic Treaty for jointly exploiting the immense mineral and marine resources of the area. The Treaty very nobly stipulates that the Antarctic area should be used only for peaceful purposes. But such exploitation for the benefit of a few technologically advanced nations can have disastrous consequences for the whole of humanity and is a matter of international concern. Christians ought to get their hands on the relevant material and begin to help raise the conscience of humanity about the need for international control of these explorations and exploitations.

Space is another source of worry. There is a great deal of debris floating around in space, and the skylab has only partially aroused public interest in the consequences of our space exploration without adequate safeguards. Much of the debris in space is military and espionage material floated by the two leading military powers. The possibility of damage to the ozone layer has yet to be computed. Only the big powers can do the computing of the damage and they may not be interested in telling us the whole story. Space, the atmosphere, the Antarctic, the ocean and the seabed are technically the common possession of humanity—areas where the notion of private property has not yet encroached. But only technically. In effect, it is only those who can afford the technology and the investment who have access to most of the World Commons, and how can the global human community have any say about what a few can do to the World Commons?

Can we keep these areas as our common property, have common control of it, and make sure that some who have technical access to it do not unduly exploit it or disrupt its contribution to the eco-balance?

We have already mentioned a number of other problems related to eco-balance, like carbon dioxide increase in the atmosphere, with the accompanying greenhouse effect, which raises the surface temperature of the earth and thereby threatens the climate and the biosphere itself.

The issue of climate change through carbon dioxide increase has been fairly thoroughly studied by a group of more than 100 experts summoned by the World Meteorological Organization (WMO) for the World Climate Conference held in February 1979, in Geneva. Their declaration puts their conclusions in somewhat guarded language:

> We can say with some confidence that the burning of fossil fuels, deforestation, and changes of land use have increased the amount of carbon dioxide in the atmosphere . . . and it appears plausible that [this] can contribute to a gradual warming of the lower atmosphere, especially at high latitudes . . . It is possible that some effects on a regional and global scale may . . . become significant before the middle of the next century.

That is a cautious conclusion and the Meteorological Conference has drawn up a World Climate Program to offset the effects of carbon dioxide increase. The biosphere, that thin and fragile layer around our planet that sustains life, has to be carefully conserved, if life is not to perish from the face of the earth. The biosphere is not even the common "property" of the whole of humanity alone. It belongs to all life on this planet. But it is now within the power of man to destroy it or to conserve it. It is humanity, and not other life, that threatens the existence of the biosphere.

Our industrial civilization is now the threat, not only to humanity's survival, but to the survival of life itself. Only when this is fully realized by humanity can a concerted effort be made to alter the course of our industrial civilization in such a way that the biosphere is repaired, healed and maintained in its ecological equilibrium.

MILITARY TECHNOLOGY AND THE NUCLEAR PERIL

We live in a time when frenzied war hysteria is being mindlessly provoked and aroused, when we hear so much about national security, European security and localized securities. We have every reason today to be concerned more about world security than about national security, and Christians face a special challenge to overcome narrow parochial loyalties and to seek global solutions.

In this spirit the WCC Conference spontaneously gave rise to a Resolution on "Science for Peace". The tone of the resolution, which was unanimously approved, is sober but unsparing. Scientists, who know well what they are talking about, appealed to all people everywhere "to accept the God-given task of using SCIENCE FOR PEACE". The Resolution says:

We, scientists, engineers, theologians and members of Christian churches from all parts of the world, participants in the WCC Conference on Faith, Science and the Future, now meeting at the Massachusetts Institute of Technology, acknowledge with penitence the part played by science in the development of weapons of mass destruction and the failure of the churches to oppose it, and now plead with the nations of the world for the reduction and eventual abolition of such weapons.

WHEREAS:

- the arsenals of tens of thousands of nuclear weapons already constitute a grave peril to humankind;

- sharp changes by the super-powers towards a counterforce strategy are so destabilizing that sober scientists estimate a nuclear holocaust is probable before the end of the century;

- there is widespread ignorance of the horrible experience of Hiroshima and Nagasaki, and the even greater implications of limited or global nuclear war with current and projected nuclear weapons;

- we are profoundly disturbed by the willingness of some scientists, engineers and corporations, with the backing of governments, to pursue profit and prestige in weapons development at the risk of an unparalleled destruction of human life;

- the waste of the increasingly scarce materials and energy resources of the world on the instruments of war means further deprivation of the poor whom we are commanded to serve;

- we grieve that so many of the most able scientists, especially the young ones, are seduced away from the nobler aspirations of science into the unwitting service of mutual destruction;

- in a time of radical readjustment of the world economy the intolerable burden of the nuclear arms race creates world-wide economic problems;

AND BECAUSE WE BELIEVE:

- that God made us and all creation;

- that He requires us to seek peace, justice and freedom, creating a world where none need fear and every life is sacred;

• that with His grace no work of faith, hope and love need seem too hard for those who trust Him;

WE NOW CALL UPON:

• all member communions of the WCC and all sister churches sending official observers, and through them each individual church and congregation;

• our fellow religionists and believers in other cultures, whether Hindu, Jewish, Buddhist or Moslem, and our Marxist colleagues;

• the science and engineering community, especially those engaged in research and development, together with professional scientific associations and trade unions;

• the governments of all nations and especially the Nuclear Powers;

• all concerned citizens of the world;

TO EMBARK IMMEDIATELY ON THE FOLLOWING TASKS:

• to support and implement the WCC Program on Disarmament and against Militarism and the Arms Race, and give special emphasis to issues related to military technology and its conversion to peaceful uses;

• to welcome and give practical support to the initiatives by the UN and its special agencies on disarmament, which affirm the right of all nations to participate in the effort to solve these global problems;

• to press for the full implementation of SALT II, to work without delay for the reduction of nuclear weapons through SALT III, and to complete at long last a Comprehensive Test Ban, all of which are urgent and necessary steps in making the Non-Proliferation Treaty effective;

• to stop the development and production of new forms and systems of nuclear weapons;

• while welcoming the exchange of scientific and technical information made possible through the Pugwash Conferences, other international scientific conferences, and the SALT process, to press for further exchanges of information as a means of reducing international mistrust;

• to educate and raise the consciousness of every constituency to the realities of nuclear war in such a way that people cease to avoid it as an issue too big to handle; in particular we recommend the formation of local study groups on the dangers of nuclear war and approaches to disarmament;

• to use every available means to restore confidence in the sisterhood and brotherhood of all, to remove fear and suspicion, to oppose hate-mongering and militarism, and to undo the policies of any with a vested interest in war;

• to prepare local and national programs for the conversion to civilian use of laboratories and factories related to military research and production, and to provide for the retraining and reemployment of those who work in them;

• to resolve never to allow science and technology to threaten the destruction of human life, and to accept the God-given task of using SCIENCE FOR PEACE.

What emerges as a basic question for the churches is this: Have we, as Christians, paid sufficient attention to what our human race is doing to our planet and its biosphere? Have Christians been too preoccupied with personal morality and social ethics in a rather limited way, ignoring the global dimension of the human impact on the environment? Humanity must grow in all three interrelated dimensions, i.e., each person must grow (in community) into God's image in holiness and righteousness; each society and all societies must become just and participatory both within itself and among themselves; but also each person and each society must become aware of the global impact of humanity upon the biosphere, and must seek justice or righteousness at the level of our relations with our environment.

And, finally, our worship and community and spirituality have to be reoriented to be faithful to all three dimensions of our existence.

CHAPTER FIVE

Science and Political Economics

ECONOMIC THEORY AS PRODUCT OF PARTICULAR CULTURES

Modern science and the technology based on it did not fall from heaven. Neither did it spring up in a vacuum. It was born in a particular cultural and socioeconomic milieu in Western Europe. It grew and developed in a specific pattern of European dominance in the world. These facts have left their marks on science technology and constitute in part their present character.

The Western science of economics also developed within that pattern of European colonialism and world domination, and our prevailing economic theory, whether neo-Keynesian or neo-classical, not only bears the marks of that pattern, but often consciously or unconsciously seeks to justify that pattern. Professor C. T. Kurien, a leading economist from India, lashed out against pretentious Western talk abut the sustainable society:

> It is a small affluent minority of the world's population that whips up a hysteria about the finite resources of the world and pleads for a conservationist ethic in the interests

of those yet to be born; it is the same group that makes an
organized effort to prevent those who now happen to be outside the gates of their affluence from coming to have even a
tolerable level of living. It does not call for a divine's (sic) insight to see what the real intentions are.

More attention must be paid to the immediate and
crying need for national and international socioeconomic
justice. The affluent nations of the world cannot discuss
the problems of over-consumption of resources and disruption of the environment without showing some interest
in putting an end to the injustice in the present world
order.

AN ECUMENICAL CRITIQUE OF CURRENT
NON-MARXIST ECONOMIC THEORY

Current economic theory as prevailing in most non-
Marxist countries seems to be a smoke-screen or an ideological cover to hide the pattern of injustice. The WCC
Conference reaffirmed the view of the Zurich Consultation
on Political Economy, Ethics and Theology (June 1978)
that "the current paradigm of political economy prevailing in Western industrialized societies, and influential in
many others" was to be criticized for:

(a) its partial perception of the humanity-nature
relationship;

(b) its bias towards the interests of a minority of the
world's people;

(c) its emphasis on accumulation and growth as the
primary answers to unemployment;

(d) its undue reliance on market mechanisms for
problem-solving and thus for achieving the greatest
good of the greatest number;

(e) its assumption that consumer demand depends
on consumer sovereignty;

(f)   its insufficient attention to the critical real world adjustment problems;

(g)   its lack of moral judgment about what is produced and who consumes how much;

(h)   its self-imposed limitations on its ability to contain the effects on economic activity of key social and political ingredients such as the role of institutions, concentrations of power and the existence of class structures.

When the Zurich document speaks of current economic theory's partial perception of the man-nature relationship, several points are implied. Economic theory regards individual human beings as producers, consumers and exchangers, and "nature" as a source or resource for commodity production. Rarely does non-Marxist economic theory take account of the fact that humanity is part of what we call nature, is integrally and inescapably related to it, and becomes human only in the process of interacting with "nature" in a social context. It regards society as composed of equally endowed and equally powerful individuals and makes little allowance for the great inequalities with which people start life, the completely lopsided concentrations of power in a few people, and the excessive power of corporations to influence the consumer's will and choice.

In the nineteenth century it was at least recognized that choices were often social, and that the economy should pursue goals other than the profit of the entrepreneur. It was in our time that an unrealistic ideology of rugged individualism and free enterprise spread in American society in order to cover up the guilt of a few individuals and groups who had "made it". Current economic theory assumes that the consumers as well as the producers are "sovereign and free" to make decisions. It assumes that the unseen hand guiding the market mechanism will create an equilibrium between supply and demand and that just distribution will take care of itself without any other social mechanism to regulate it. While it has recognized the power of government to introduce certain regulations in the patterns of production and distribution, it does not take sufficient account of the fact that this government itself can be controlled and manipulated by the corporations and other vested interests. It assumes that if overall production keeps growing, the problems of unemployment and inflation will take care of themselves. It does not take into account the power of corporations to exploit the labor and resources of other countries and to impoverish them. It does not recognize that international banking and financing are themselves means of exploitation and oppression. It does not recognize the power of corporations, farming lobbies, and their money to influence public policy not only in the developed countries, but also in the less developed countries. It does not acknowledge the fact that its mechanisms are calculated to assure the welfare of a few at the cost of the many.

But these are not its only defects. It contains subtle ideological elements which corrupt the lives and values of people in all countries. It assumes that more is better and seldom recognizes that enough is enough. It contains no stipulation for deciding what commodites shall be produced and who shall consume them. It too easily identifies need with demand, and does not recognize that even need can be created by propaganda and pandering. It does not always distinguish between the basic needs of all and the luxury demands of the pampered few.

Worst of all, it creates an ideology for worldwide consumption in which the affluent society sets itself up as a model for others to emulate and imitate and catch up with. By creating distinctions like rich and poor nations, economic theory lays it down with great subtlety and deceptiveness that the task of the poor is to become rich, the task of the less developed is to become more developed; thus economic theory, posing as science, becomes a major tool of mental as well as economic enslavement. A worldwide pattern of dependence is created in which the affluent regard themselves as the center and the "less developed" nations as the periphery—the developed First World and the Third World which looks up to it for intellectual guidance, financial aid, technological assistance and cultural norms, so that it can "catch up" with the "developed" world. It is this pattern of mental, spiritual, cultural and economic enslavement that is reinforced with the aid of Western economic theory masquerading as "science", and which the gullible intellectuals of the Two-third World so readily swallow from Western textbooks and institutions of higher learning and propagate it in their own societies.

WESTERN ECONOMIC THEORY—A LAY COMMENT

Even the physical sciences are marked by the culture in which they rose, as we will see in a later chapter. The social sciences are bound to be more so. Whether it is the economic theories of Aristotle in pre-Christian Greece, or of Kautilya or Chanakya in Ancient India, the values and norms are largely taken from current society as well as from the models chosen in terms of which one explains economic activity.

What now passes for scientific economic theory has its origins in colonial England. All of the three main lines of economic thought—the classical theories of Adam Smith, David Ricardo and John Stuart Mill, the counter-theories of Marxist thought, and the "volcanic eruption" of John Maynard Keynes, can be traced back to an imperialist Britain in the throes of the Industrial Revolution. Of course there have been many footnotes—the neo-classicism which is still in vogue, the neo-Keynesianism which seeks to wed classical economics to the critique of Keynes, and the neo-colonialist economic theories of Walt Rostow and Daniel Bell, but these are only footnotes and readjustments in the light of the interplay between the three sets of theories and the anomalies of economic reality today.

The task of economic theory is always a twofold one: (a) that of creating models or simplified paradigms of relationships between calculable or measurable variables considered most important in the analysis of economic activity as we observe it, including their empirical verification in relation to the realities of history; and (b) the concomitant process of setting up norms, laws and orientations for what needs to be done in order to keep the economy healthy, and for deciding on the mechanisms needed to guide the economy in the desired direction.

Economics, as a science, then, is both descriptive and normative. The descriptive element seeks to articulate the "laws" of demonstrable regularities and recurrences of economic events; the normative element gives direction to governments, corporations, trade unions and the general public in matters of taxation, wage and price control, and the fostering of institutions and processes.

The descriptive element in classical economics saw the elements of production as threefold—land, capital and labor. This theory arose in the midst of conflict between the interests of a landed aristocracy and a new class of traders and entrepreneurs competing for a larger share in the fruits of the exploitation of the agricultural and industrial working class, both at home and abroad in the colonies.

The entrepreneurial class was interested in the development of science and technology, both as a means of increasing production, and therefore profit, and as a way to reduce the dependence on human labor, which was becoming more and more expensive as wages and standards of living kept rising. Machinery and labor, according to Ricardo, are in constant competiton, and the development of machinery through science-technology would be the defense of the capitalist class against the specter of rising wages and the strident demands of the working class.

But the development of machinery requires the accumulation of capital, which can be done only by keeping back some of the fruits of labor from the laborer. The laborer needs to be maintained at a certain minimum level of health, nutrition and basic need satisfaction, and the closer to the minimum his wages are, the larger the profit margin, which can then be used for further capital development in the form of machinery for production. And the more capital there is, the greater the productivity of the laborer, who can thus be given a slightly higher wage, keeping in mind the need for the maximization of profit and the expansion of capital equipment, which, along with land (the source of raw materials) and labor, constitutes the main elements of an economy.

The development of science and technology was thus from the beginning wedded to the need for the capitalist to increase his profit margin, though it also paid dividends to the labor class. The trade unions sought to maximize these dividends, while it was in the interest of the capitalist to minimize the laborer's dividends and to maximize his own profit margin.

Both the development of science and technology and the development of economic theory took place in the context of this conflict of class interests. The capitalist class had to fight it out on both fronts. On the one hand there was the feudal or landowning class, who, remaining idle and unproductive, collected rents and prices for the raw materials produced by the land. As wages and prices rose in consequence of the increase in capital equipment and its qualitative improvement through science/technology, the landlord reaped a portion of the benefits by collecting higher land rents* and raw material prices. This fact meant that the landlord cut the profit margin of the capitalist. Thus the feudal aristocracy and the industrial laboring class clamoring for higher wages were inimical to the interests of the new class of entrepreneurs.

Even Marxist thought grew up in this milieu, representing the aspirations of the agricultural and industrial laborers. Karl Marx developed the two related theories of "value" as created by labor alone, and capital as "surplus value" kept away from the workers. According to Marx, labor was the single source of value. Marx did not mean that the value of each commodity was the amount of labor that went into it. That may have been what the classical economists thought. Adam Smith, for example, held that two things requiring the same amount of manufacturing time would have the same value and should sell for the same price.

*The increase in population also led to pressure on land, hence to higher rents, hence to higher wages.

For Marx however it was the sum total of socially organized labor in a given community that created value. All the goods and services produced by a community together constitute the fruit of the labor of the total community. Each commodity is the materialization of a given portion of that total labor, and that portion constitutes its real value, irrespective of what anybody pays for it. Its exchange value in the market, as well as the wage paid to the laborer, can both be different from its real value in terms of the portion of social labor that has gone into it.

Now for Marx, the community that produces the commodities has at its disposal many other things—resources, land, machinery, technological know-how, and so on. And the community labor uses all these things to produce commodities, but it is labor that constitutes value. Given all the other elements, it is labor that makes them into commodities for use. And the whole value of the commodities belongs to the community that created them. In the capitalist economy, says Marx, only a part of the value created is paid back to the worker as wages, the remainder, or the surplus value being held back by the capitalist as his profit.

Mathematically expressed, the total value of the social product (SP) is equal to the sum of three variables CCVCSV. CC stands for constant capital expenditure (i.e., depreciation of equipment, consumed raw materials, energy, etc.); VC stands for variable capital and means primarily the sum total of wages paid to the laborer; SV stands for the surplus value appropriated by the capitalist. And if MP stands for market price, we can say:

$$SP=CC+VC+SV$$
$$\text{but}\quad SP=MP-(CC+VC)$$
$$\text{and the profit rate}=\frac{SV}{CC+VC}$$

But this profit rate has to be multiplied by the speed of turnover of capital. The faster the turnover, the higher the rate of profit. So the capitalist's interest is (a) to increase the surplus value and (b) to quicken the pace of turnover of capital.

Now part of the surplus value is utilized by the capitalist for his own consumption and what remains goes into capital accumulation and improvement of the efficiency of production. The ratio between wages paid and capital expenditure is important. If by employing the same amount of labor but using more efficient capital equipment the social produce can be increased, then, for each unit produced there is more profit. Furthermore, if the total amount of wages paid is less, the surplus value increases again. So the capitalist has two interests: (a) make the capital equipment technically more and more efficient, and (b) reduce as much as possible the amount of total wages paid for labor (by reducing the number of employees).

On account of (a) the capitalist is interested in the development of science and technology insofar as these can help him to produce more goods per unit of capital invested and labor used. But on account of (b) too, the entrepreneur is interested in science and technology insofar as they can reduce his dependence on human labor by relying more on mechanical automation.

One of the main sources for funding scientific-technological research is, of course, the large corporations. Their interest, however, is largely in science/technology for a particular purpose—increasing cost-efficiency, margin of profit and dependence on human labor. This means science/technology receives an impetus in a market-economy society towards the development of production machinery. It may also slightly improve the quality of the product the consumer receives. But by and large the benefits of science and technology of this kind ensues to the entrepreneur or the corporation, rather than to the consumer or the laborer.

The other major source of funding for research is in the military establishment. Here again the direction of development in science and technology is clearly anti-human, in the sense that the ingenuity of man and the resources of the earth are used up not for augmentation of the quality of human life, but for mutual destruction and inhumanity. The UN Conference on Science and Technology for Development, held in Vienna in 1979 pointed out that 50% of present science and technology is now in the service of military establishments, and that half a million of the world's scientists and technicians are on military pay.

Who makes these decisions about how our human and natural resources are to be utilized? Clearly the people do not want their energies and resources wasted on war. What unseen hands and heads are responsible for these decisions which everyone recognizes as partly irrational and partly mad? Can economic theory also deal with such questions? Does the one-man one-vote principle ensure equality in decision-making power?

In the nonsocialist countries, political science and economics are taught as separate disciplines so that the influence of politics in economic decisions and of economic interests in political decisions are not very clear even to the political scientist and the economists. Even India has few real political economists involved in planning and policy-making structures. The result is that whatever economic development there is works more in favor of the already privileged.

If the people's awareness of these issues is to be heightened, there has to be a radical change in the teaching of economics in our universities. At present economics as science seems to be distorted and used against the interests of the underprivileged. Whether it can be called science is a question economic and political theoreticians and philosophers should pursue with urgency.

TOWARDS A NEW SCIENTIFIC POLITICAL ECOMOMICS

Economics, or rather political economics, is the basic science which should deal with ensuring the human dignity and human worth of everyone in society, regionally, nationally and globally. Its concern cannot be limited to the mere production and distribution of commodities.

Political economic theory must provide for the following elements to be adequately dealt with:

(a)   it must constantly formulate, reflect upon and reformulate, on the basis of a social consensus, a set of
values or social goals for which production is organized;

(b)   it must formulate and constantly reflect upon the
principles of sociopolitical organization for production,
distribution and decision-making with a view to enhancing justice, freedom and human dignity for all;

(c)   it must provide for full participation of all sectors
of society in the decision-making structures of the
politico-economic organization at all stages: planning,
execution, and evaluation;

(d)   it must deal effectively with the questions raised
by the human impact on our environment and on the
biosphere;

(e)   it must provide for the conservation and augmentation of the best in all cultures, and for their greater
development in a more humane direction, including the
development of religious faiths and institutions;

(f)   it must provide for interregional, intercultural
and international relations which promote unity, mutual understanding, justice and peace in the whole
of humanity.

This means that a new science of political economics
should also deal with the cultural and religious heritages
of humanity which have, *pace* Marxists, been often more
decisive for human development than merely the social
relations of production and distribution. In the nineteenth
century when Karl Marx and Friedrich Engels wrote,
religion was regarded as a reactionary force retarding the
growth and development of humanity and supporting the
vested interests of the establishment. This judgment of
nineteenth century European religion may or may not
have been valid. But there is no reason to take that judgment, limited to one particular time and space in the his-

tory of global human development, as a universal principle.
The Western scientific ethos is secular, and it shies
away from dealing with religion as an integral part of the
political economy. But does science have to be necessarily
set in a secular framework? True, the secular frame provided for the free growth of Western science, unfettered
by the shackles of ecclesiastical authority. But has not
science today sufficiently come of age to declare its own
independence from the secular bind that is by no means
essential to it?

Christians have a responsibility to pioneer here. The
MIT Conference showed very clearly that there is no essential conflict between modern science and the Christian faith, or for that matter between science and other
religions. It is possible, not only to accommodate all the
valid insights and discoveries of modern science within
a religious perspective, but even to give new and more
creative orientation to science and technology by integrating them into a conceptual framework of creation-
incarnation-eschatological fulfillment.

A Christian scientific political economics is not a contradiction in terms. Its basic structure can be delineated
only after some of the philosophical issues relating to
science and to our human relationship to creator and creation have been more adequately set forth.

CHAPTER SIX

Science and Philosophy

The characteristic feature of Western academic knowledge today can be seen as a mixture of extreme specialization on the one hand and the desire for interdisciplinary and integrative studies on the other.

The integration and interrelating of the various sciences still remains a fairly impossible task, primarily due to the recent explosion in the quantum of available knowledge. No human mind can assimilate more than a small portion of the main aspects of knowledge available. In spite of great advances in cybernetics or in information storage and retrieval technology, we do not yet know how to program any real integrative techniques into the computer.

We have no reason to be impatient with science. After all, in its present form it is less than two centuries old. The first university degree in science in the U.S.A. was granted only in 1865.[1] And ninety percent of all the scientists the world has ever produced are living now. For example, we have a million chemists in our world today; every year we produce some three million scientific papers. Modern science is still in its youth, rather prodigious in its energies, though perhaps still far away from emotional and integrative maturity.

Science, because it is so young and so vigorous, so impressive in its performance, gets invested in our culture with a sort of guarded mystique. Nobody openly says science is omnipotent, but underneath many people assume it is. It is the "open sesame" to all knowledge, the magic password for entry to the human or other planetary spheres or into the depths of the atom.

MODERN SCIENCE—ASSUMPTIONS AND IMAGES

But every culturally conditioned variety of science such as ours has its own unscientific assumptions, many of them somewhat unexamined. Just to mention a few, it has a cosmology or world view—a picture or mathematical equation of how the universe is constituted, what processes, mechanisms and structures govern its functioning, etc. Some very learned books, like Weinberg's *The First Three Minutes*, describe the whole process of creation as it emerged in the first three minutes of its existence in a manner that leads one to believe that this reputed scientist is simply describing reality as it happened, not extrapolating from our present sketchy knowledge of the structure of our universe, to what must have happened "in the beginning".

Modern science also implies many unexamined assumptions about the nature of man. The scientist in general, and not merely the biologist or the anthropologist, works with certain conceptions of what human possibilities, functions and faculties are. In that process he manages to ignore a vast amount of data which point to other human possibilities and functions—the paranormal phenomena of ESP and so on, the religious experience, the drug experience, the experiences of love and joy, and the fear of the Lord.

Modern science has also a set of value assumptions drawn largely from the culture in which it grows up—e.g., that knowledge can be made into a commodity, stored, traded in, copyrighted, made into the property of an individual or a group; or that knowledge is good in itself; that the knower is distinct and separable from the known; that the world open to our senses exists "out there", independently of our perception.[2]

Modern science entertains also a set of epistemological assumptions—about the knowability of the world, the homogeneity of nature, about the subject-object dichotomy and so on; that errors can be tested and overcome by experimentation and rational reflection.

(a)   *The Platonic Image of Science:* Its basic inspiration seems to be the platonic one—the assumption that some specially gifted and trained people can, by the deductive method based on experience and dialogical reflection:

(1)   know the truth;

(2)   restructure the world in perception with reference to the real and unchanging ideas or natural laws;

(3)   overcome errors in perception or theory by rational reflection and rational action, undertaken and led by the disciplined modern philosopher-kinds or scientists.[3]

Despite the antiplatonic and pro-Aristotelian bias of much of modern science, there is still much truth in Professor A. N. Whitehead's dictum that most of Western civilization is but a series of footnotes to Plato.

(b)   *The Pathology Image of Science:* But the scars of the battle between the rational enlightenment and ecclesiastical authority in modern times are still visible in Western culture—and not only among religious fundamentalists who distrust scholarship and rational reflection. Jacques Maritain, the famous neo-Thomist philosopher of our times, spoke about the "Pathology Image" of modern science in our culture. Modern science could be conceived as a "deadly disease" alienating man from reality, eroding human faith in moral absolutes, inimical to the cherished values of the so-called Judaeo-Christian faith which people claim as the foundation of Western civilization.[4]

(c)   *The Model Image of Science:* In the Anglo-Saxon world, however, the philosophical understanding of science has moved from its earlier positivistic assumptions to a more modest "model" image. Science is a machine, a tool, which enables us to do certain things which we otherwise could not; no ultimate truths can be deduced from science; that is not its function. As Professor Mary Hesse of Cambridge expounds it, science can be understood on the model of a teaching machine with feedback control. Out of our present culture and language we develop theories which we think will fit "nature", and also empirical experiments which are theory testors. The result of the experiment helps us to correct our theories, and newer and more apt hypotheses are then put to the appropriate new experiments or theory testors. By constant feedback, theories are constantly improved, and the process yields us operationally useful knowledge, and some of the regularities observed may also be indications of a truth that goes beyond the operational.

This pragmatic, operational view of scientific knowledge is indeed a far cry from the positivistic claims of an Auguste Comte or an earlier crop of Vienna circle philosophers and scientists. They no longer talk of the "laws of nature", but only of "law-like statements" which are operationally useful. The "model" image is a modest image.

(d)   *The Symbolic Form Image:* What is becoming increasingly fashionable in Western philosophy of science seems to be the neo-Kantian symbolic form image. Ernst Cassirer and his disciple Susan Langer introduced these concepts long before the philosophers of science got hold of it.

Immanuel Kant had already proposed that the human mind is not a passive *tabula rasa* which receives impressions from the outside world through sense experience, but is an active cocreator of knowledge or concepts, through the forms and categories supplies by the structure of the mind itself for the formation and interpretation of experience.

Kant's error was perhaps in attributing to the universal human mind the characteristic structure of the eighteenth century German mind putting into that structure Newtonian mechanics, Euclidean geometry, and Aristotelian notions like cause and substance, giving these an *a priori* character. Some of that tendency is still seen in cognitive psychologists or "generative grammarians" like Noam Chomsky, who for example argues for a "fundamental grammar" or "deep grammar" for the human mind, a grammar which is given, universal and rule-conforming.

But today many philosophers of science argue that modern science is but one of the possible ways of perceiving the world, other ways being, for example that of art and poetry, or religion and mysticism.

TOWARDS A DOMINANT IMAGE WHERE DOMINATION IS NOT CENTRAL

Of the four images, the most popular among educated people is the optimistic platonic image, though many pious people do still retain alongside the pathology image. There is a tendency among pious people even to question the adventures of science as the fruit of unlicensed human pride. We hear this every time there is a landing on the moon, a splitting of the atom, or the fertilization of a human embryo in a glass dish. The charge is that man in his pride is usurping the place of God, and will soon be chastised by a jealous God who does not want his special prerogatives to be taken over by man.

The coexistence in our cultures of the platonic image and the pathology image of science lies at the root of our ambivalence towards science—on the one hand coveting and desiring it as the means to the solution of all problems, and on the other mistrusting science as capable of bringing down the wrath of God upon us.

The model image is in a sense a modest pose claiming to make no value judgments about science except that it is operationally useful. It fits in very well with the pragmatic tempo of the Anglo-Saxon world and its basic antiphilosophical, antimetaphysical attitudes. But the Anglo-Saxon thinkers give us little clue as to what science is all about. If it is only an operational tool, one among many others, then why does it come to have such influence in our societies, and rule so despotically in our academic communities? Why is it still true that the man in the white coat, operating the buttons of a computer, still fills us with some kind of awe? There must be some truth to the allegation that science/technology has replaced religion as the source of authority in our urban-industrial culture, and the scientist in the white coat is now a surrogate of the priest in the black cassock. And it would not do therefore to dismiss the fundamental questions about scientific knowledge with the simple assertion that science/technology is but a tool and leave it at that.

If, as Rosemary Reuther claims, both our modern science and the theological matrix which produced it were creations of a male-dominated culture to which the concept of domination is the *key*, we need then to ask ourselves the question whether we can develop another kind of science/technology, and another kind of theology in which love, joy, peace, compassion, and kindness, rather than domination and manipulation, provide the central ethos.

Here, theology also needs to be radically reformed. For the element of domination-manipulation is still too central in theology. Even in supposedly antidomination theologies like Black Theology, Feminine Theology and Liberation Theology, one can very well hear, if not the rumbling desire of the hitherto dominated group to dominate their dominators in retaliation, at least a basic lack of love and compassion, or an absence of joy and peace.

If the element of domination-manipulation seems thus common to both theology/social ethics and science/technology, then clearly it is a reflection of the way society itself is organized, and we are hardly likely to arrive at a theology/social ethics or a science/technology that is peaceful and joyful, until some necessary changes have taken place in the structure of society itself.

But the transformation of science and theology should mark the transition towards a peaceful and joyful human existence, with love and compassion. Here the conceptual must go hand in hand with the sociocultural and politico-economic transformation. But what has to be resisted is the temptation to short-cut the passage to the sociocultural and politico-economical, without going through the conceptual. To deal with only the ethical issues posed by science is to treat the symptoms without diagnosing the cause of the disease—a temptation to which Two-third World thinkers also too often succumb. For they, too, are educated and formed by our pragmatically oriented, science-and-technology-based civilization. That is why we must do more work on the image or conceptual construct of science, as well as on the underlying perception of reality itself, if we are to deal adequately with the ethical issues posed.

Here philosophy comes in as a necessary tool in analysis. Our academic communities, however, as a result of the domination of science and technology in the university structure, have lost their capacity for clearer philosophical reflection.

The integration of human knowledge, theoretically at least, remains the province of the academic community, which has access to that knowledge, even though in fragments. But the academic community is precisely the place where deeper philosophical reflection seems to be generally discouraged.

When philosophy fails to be faithful to its own true vocation, the integration of knowledge becomes an impossible task; we resign ourselves to a mere socioeconomic or ethical analysis of our problems. In theology, as well as in social ethics, we take the view that the problem is one of political economic analysis and reconstruction.

Philosophy, at least academic philosophy, has abdicated the task of integrating our vision of reality and providing a coherent interpretation of reality and of our relation to it. Today it is content with an analysis and criticism of received social, scientific and religious propositions, and the reconstruction of principles and categories regarded as indispensable to correct theories or sound policies of action.

We have then to proceed through an analysis of philosophy and theology, to see what their true function is, and how to make philosophy functionally effective. Before we do that, we should perhaps engage in some reflection on our ways of ethical decision-making. We may be in a better position after that to engage in further reflection on the present states of the philosophy of science, of philosophy in general, of Christian theology and perhaps of the religious approach to reality as distinct from the secular approach.

To find a science based on compassion, peace and joy, demands a gigantic effort of the imagination. Herein lies the great task of the Church in reorienting the development of science/technology away from domination to a compassionate service of humanity.

CHAPTER SEVEN

How Does One Decide?

THE NATURE OF ETHICAL REFLECTION AND DECISION

The Christian churches need to ask themselves five questions:

(1) How in fact do people arrive at personal, family or social ethical decisions?

(2) How does faith contribute to ethical decisions? Does science contribute?

(3) When there are different theologies—can you still come to an ethical consensus?

(4) How does one translate theological ethics into secular ethics?

(5) Should the churches take the lead in promoting free enquiry or should they make up their minds and take partisan or advocacy roles? Should the churches legislate for their members on ethical questions?

THE PROCESS OF ETHICAL DECISION-MAKING

Professor Karen Lebacqx of the Pacific School of Religion has posed the question in a fresh and interesting way. In reviewing the contemporary methods of ethical reflection, she points out the limitations in our present approach, which she characterizes as too decision-oriented, too individualistic, too a-historical.

In decision-oriented ethics there is the assumption that there is always one right thing to do in a particular case, and that this can be found out by a rational analysis of the issues and principles involved. For example, in the case of a mother told by her physician that her four-month-old embryo in the womb is genetically deficient (say Down's syndrome or Mongoloid disease), should she agree to an abortion? One could argue that the child when born is going to suffer and cause suffering to others, and therefore that it is better to prevent the birth of such a child. Or one could calculate how much it would cost to bring up such a child for the twenty to twenty-five years of its life and what contribution such a child or young person could make to family and society. In either case, the attempt is to decide the issue on the basis of the degree of suffering or happiness, or of economic costs and benefits. The more fundamental question, whether one can justify the destruction of a four-month old human embryo, is sometimes discussed, sometimes answered, with the general principle that abortion of human embryos is justified under certain conditions.

There are other hidden assumptions behind such a process of ethical reflection. Some people seem to think that it is possible for human beings always to do the ethically right thing and thus be sinless and unguilty. Others would say that any human action is bound to be sinful and that one can only be justified by faith.

In all these considerations, one has a basically individualistic orientation. Especially in the developed countries such decisions are taken in the context of a physician-patient relationship, sometimes a religious leader or pastor joining in the discussion. Karen Lebacqx has said that the famous "patient-physician relationship", i.e., the luxury of having a family physician with whom one can discuss and decide on ethical issues, is available to less than two percent of the population in most of the countries of the world, and that this way of ethical reflection and decision is not one of the options that they have. In such countries, the facilitation of health delivery to all sectors of the population becomes the basic ethical issue and this is not something on which the individual can reflect and decide. But it remains a fact that particular individualistic ethical dilemmas are a luxury of the richer classes and do not carry much interest for the majority of the people of the world who are too poor to have any kind of medical assistance at all.

There is also the assumption that scientific data are totally value-free. If a doctor says to an expectant mother that there is a fifty percent chance that her child would be genetically defective, such a statement can be shown to be value-laden at several points. In the first place, one could have also said that there is a fifty percent chance that the baby would be normal, which would be another way of stating the same thing. Equally value-laden is the word "defective" as we have already pointed out. The doctor could have stated the same thing in other less value-laden words.

Karen Lebacqx believes that the women's liberation movement is discovering better methods of ethical reflection and decision-making. For example, according to her, women prefer reflection about structures of society and patterns of meaning rather than rational analysis of personal ethical dilemmas. This may not, however, be an either/or affair. Perhaps structural thinking may provide a framework for resolving ethical dilemmas; but perhaps they may not. It is more likely that we have to deal with social structures as well as with personal dilemmas in ethical reflection, and only in some cases would the two be clearly interrelated.

Professor Lebacqx also maintains that in liberation theology the emphasis even in social ethics is changing.

At one time the call was for a prophetic passion for social justice; today there is a shift to the apocalyptic, i.e., the attempt is to get at the name and number of the beast (Babylon—666, in the biblical book of Revelation) which is behind all the abominations in society.

The women's movement, according to Lebacqx, is making a major contribution in its insistence that it is not sufficient to analyze the *sociopolitical* structures but that one should go deeper into the *thought-structures* and patterns of meaning that underlie people's aspirations. Ethical questions cannot be resolved at the ethical level; they have to be transformed into theological and philosophical questions.

A clear example of this would be, in the Indian context, to ask the ethical question whether it is right to give alms to a hungry beggar. Sophisticated Indians would say that we cannot solve the problem by giving alms and that we should create the kind of society where there would be no beggars. Well and good, but what about the present hunger of the poor man? Can he live on the pleasant thought of a future society without beggars? Behind the typical sophisticated Indian's refusal to give alms to a hungry beggar, there may be several concealed thoughts:

(a)   Perhaps this beggar has a bank account* and by giving him alms I am being reduced to a fool who contributes to somebody else's bank account;

(b)   If I give to this beggar, others may also approach me and I do not want to be the one taxed for being compassionate in one instance;

(c)   After all, I earn my money by hard work; why should I give any of that to this unproductive parasite on society?

(d)   I pay my taxes to the municipality. It is their job to do something about those beggars, both to provide

*Many professional beggars in India have been found to have bank accounts or significant sums of money otherwise stored away.

for their living and working, and to make sure that they do not pester honest citizens;

(e)   Charity is dehumanizing. What we need is justice;

(f)   What can I do by myself to solve this problem? I do my duty by doing honest work. It is up to society to do something about it;

(g)   What I have is mine. Why should I give it to others?

(h)   Begging is exploiting. I refuse to be exploited.

We will not attempt here to analyze these thoughts, but merely point out that there are thought structures behind the ethical decision not to give alms to a hungry beggar. It is the task of philosophy and theology to get behind the ethical issues and tackle the thought-structures of people and the values implied.

According to Karen Lebacqx, women's movements are resorting more and more to story-telling ("her story" rather than "history") as a way of ethical reflection and decision. When confronted with the ethical dilemma of a woman facing the issue of whether to have an abortion or not, we should ask the question: What is the story of this woman? What is her life situation? In the Black liberation movement as well as in the women's liberation movement, life-stories of persons and groups have had a more telling impact on social action and ethical motivation than cold and abstract analyses of social and economic structures.

Once again, it seems that it is not a question of either/ or. We need to do all these things:

(a)   listen to the stories of the oppressed, the downtrodden and the marginalized, and retell these stories to others;

(b)   undertake adequate analyses of the sociocultural and political economic structures which cause the oppression and exploitation;

(c)   go even further back to theological and philosophical analysis of the thought structures and patterns of meaning lying behind the socioeconomic structures;

(d)   deal both with personal dilemmas and with what would constitute long-term solutions.

FAITH AND SCIENCE IN ETHICAL REFLECTION

The role of science in Christian ethical reflection is easier to delineate than that of faith. For faith remains a nebulous concept, even after making the distinction customary in Western circles between *fides quo* and *fides quas*, "belief in" and "believe that". Faith is more than trust in a person or believing that certain things are true.

The WCC Conference's Section I was devoted to the Nature of Science and the Nature of Faith, with which topics we shall deal in a later chapter. The basic stance of that document is Western Protestant and it set out the dilemma of Western Protestant contextualist or "biblical" ethics:

> We struggle to find an ethic more secure and authoritative than our feelings and our social location. If, for example, we appeal to conscience, we find that our consciences are largely determined by our societal experiences. If we search the Scriptures, we find that the parts that move us most powerfully are those that address us where we are, that the concepts by which we interpret the Scriptures are those that we have developed in a given historical context.

Section X, defining faith as "a response to God, a process, a directing of life, which influences the whole person and the Christian community,"saw "Christian behavior" as "rooted in the Christian love and understanding of God." They took the line that the heart of this understanding is an anticipation and celebration of the "new era for humanity which God inaugurated in Christ".

Without basically disagreeing with that position, an Eastern Orthodox theologian would like to put together two concepts to amplify that view—the concepts of Baptism-Eucharist and *Theosis*, as the basic framework for Christian ethics. These framework concepts are both eschatological, i.e., they involve participation in the new age as well as anticipation of its fulfillment beyond history. Participation—not merely celebration and anticipation or a mere pointing to the messianic kingdom—is the key for Eastern Orthodox understanding of the new age. The new age is not a mere intellectual construct that provides insights; it is an actual living of the new life inaugurated in Christ's death and resurrection.

Baptism is the initiating mystery, by which through faith and the action of the Holy Spirit, Christians are incorporated into the reality of the new age through participation in the death and resurrection of Christ. One now becomes a member of the historical-transcendent community of faith that experiences the new life and grows in it. The Eucharist is the mystery through which Christ the High Priest perpetually offers up himself along with His Body, by the Holy Spirit to the Father. Baptism-Eucharist is thus the actual process of participation in the risen life of Christ the God-Man who unites the community of the Church with the community of the Holy Trinity.

This existence in the Church and in the Holy Trinity is a process with marked "ethical" consequences. The process is called in Eastern Orthodox theology by the name *Theosis*. *Theosis* means the progressive separation from evil and advancement in the good; humanity being created in the image of God should be able to become more like the original, i.e., God. Only the character of God as infinite good is the normative limit for growth in the good. And since that good is not finite, there can be no stopping in the good; there is no end point where one can stop growing. But man being finite, there is no risk that the addition of any amount of finite good will make him infinite like the original. There can be only approaching infinity, but never achieved infinity. But as you approach infinity, you also realize that there cannot be many infinities.

This process, however, takes place now in a world where sin is integrally woven into every form of good, and no form of good ever remains static if it is to remain good.

Hence the struggle against sin is bound to be perpetual in history. There will be no dawn in history when sin and evil will have completely disappeared.

Hence the constant and unrelenting struggle against evil, both personal and social, must remain an integral aspect of historical human existence. Only perpetual vigilance can keep the good from turning into evil. The Christian thus entertains no vision of a "classless society" in history in which injustice, oppression and exploitation have been permanently banished and people will live happily ever after. The separation from evil takes place only at the end of history, through death and resurrection, through the transformation of the body and its way of perceiving and dealing with reality. This does not mean that the body is responsible for the presence of evil. It merely means that the present body is the principle of historical existence, and so long as we are in this perishable body we will also be in historical reality, where the wheat and the tares always grow together.

This does not, however, mean an acceptance of the inevitability of evil. On the contrary, since evil has been in principle overcome in Christ, it is our job to continue that war without fearing the power of evil, to be vigilant and watchful for new forms in which evil will appear, and not to be fooled by evil appearing as good, or to be blind to its presence in ourselves and in groups and institutions with which we are identified.

The negative struggle against evil is only one side of the process of *Theosis*. The other side, which has to be simultaneous, is growth in the good, creation of the good, promotion of the good, hungering and thirsting for righteousness and holiness, in oneself as well as in the whole of humanity. The gifts of the Holy Spirit are available to us in the community of the spirit precisely for this bearing fruit in love, joy, peace, self-control and heroically creative good. The growth in the good is also a community process, the center of the process being participation by the community in the death and resurrection of Christ in the Eucharist, "until I Come", that is until the end of history. In the Eucharist, the Church offers up herself and vicariously the whole of humanity as well as the rest of creation, in Christ, by the Spirit, to the Father; and receives from the Holy Trinity the divine life through the medium of the body and blood of Christ. It is this divine life which then has to be lived out in the midst of history, bearing fruit in creative good.

For the Eastern Orthodox, one cannot jump directly from the Bible, or from the situation, into ethical issues; rational discussion and ethical decision-making form but part of the business of eucharistic existence. It is not what one thinks or does that provides the foundation, but one's consciousness and spontaneous creativity as they are formed through the process of *Theosis*.

This perspective on how faith affects Christian ethics is one that needs to be seriously explored in an ecumenical context.

As for science and its contribution to ethics, the report of Section X does mention two kinds of contribution: (a) the values inherent in the scientific enterprise itself, namely "honesty, a humility before truth and a willingness to set aside prejudice and accept correction from evidence", and (b) "freedom to search for truth". It recognizes that these values are not sufficient for the guiding of the scientific enterprise, and that humanistic values have to undergird science policy or the development and utilization of science and technology.

Scientific data in themselves do not yield a science policy. However, "given certain assumptions about values and purposes, the data may point to a policy".

Technology is more closely related to ethics—making possible the cure of disease, the production of more food, swift communication, easier exploration of surrounding reality; but also invasion of privacy, greater exploitation and oppression, greater mutual destruction, quicker destruction of the biosphere and so on.

Technology is power, based on science as knowledge. And all power is ambiguous, capable of use in the service of good or evil. Technological advance thus creates the possibility of increasing the power of evil or increasing the power of the good. The fact that fifty percent of our scientific-technological power is now in the service of war and destruction, and a good portion of the other half in the service of quick profit or imperial expansion for the corporations is merely an indication of whom science and technology now serve. The ethical issue then becomes not merely that of dealing with personal moral dilemmas produced by new technological possibilities. The main issue is that of liberating science-technology from the bondage to evil and injustice and war.

Eastern Orthodox theology would go further. The Eastern patristic view is that man becomes fully human in learning to coordinate head and hand, both being controlled by the heart, which is the center of one's being, which in turn is guided and directed by the spirit of God in community. Science-technology is a sort of head-hand coordination, and leads humanity to greater maturity, and complexity of personality and society, as well as conceivably human brain evolution. The actual failure of Eastern Orthodox theological reflection in recent centuries has been the failure to take this seriously. There have been but few Orthodox thinkers who have adequately studied the complexities of modern scientific technological civilization and then proceeded to write Orthodox theology.

In any case the Eastern patristic tradition would not be negative in its attitude to the development of science and technology, but would, on the contrary, encourage science-technology as a necessary development in the growth of historical man in process of *Theosis*. The assimilation and control of science and technology would be part of the way humanity grows—in Christ, no less.

If this be so, Eastern Orthodox theology cannot revert to any lazy romanticism which wants to backtrack to a pretechnological era to find peace and tranquility. We must go through this process of scientific technological development, but keeping two things in mind: (a) it is not a final stage, where the universe yields all its mystery to human curiosity through science-technology, but it merely opens up one aspect of reality in such a way that the human capacity for creation of good and evil is enormously enhanced; and (b) it is a knowledge and skill which have to be mastered and brought under control before they over-whelm us and destroy us.

In other words, Eastern Orthodox theology would take a positive attitude towards science/technology without being overimpressed or mesmerized by it. It is one way of dealing with reality; it provides immense possibilities for the creation of the good; it is a head-hand coordination skill which we have to acquire in the process of the evolution of the human race; it should be adequately brought under social control, so that it really becomes a tool for the creation of the good in the hands of the whole of humanity and not just a privileged few. But it should not be made the sole way of knowing, and it does not lead, in any case, to any ultimate verities. It enhances human power to create good or evil. Divorced from love and wisdom, science/technology becomes an enemy of humanity. Because it gives more power, it has to be carefully watched, so that the additional power does not serve the interests of injustice, oppression and exploitation. Head-hand coordination should be further coordinated with growth in the good. Thus both science and faith should be at the service of the good—power at the service of wisdom and love for increased creativity in the good.

THE PROBLEM OF DIFFERENT THEOLOGIES

The WCC Conference consciously planned to have more addresses by scientists than by theologians. And hence the problem of different theological starting points was not at the center of the debate. There were theological addresses on The Nature of Faith (Gregorios), on Humanity, Nature and God (Birch, Liedke, Borovoy), on The Christian Approach to Science and Technology (Falcke), and on Bio-ethics from a Liberation Perspective (Lebacqx). The methodological problem of arriving at a common ethics in a world of many theologies and ideologies remains still to be tackled. Our horizons were stretched by the addition of several addresses from the perspective of other religions and ideologies.

With different starting points one can come to similar conclusions, we discovered. Quite often the different perspectives complemented and completed each other. The perspective was substantially affected also by the socioeconomic situation from which the speaker came. For example, while a speaker from an industrially advanced country would tackle the problem of resources from the perspective of being finite, a Two-third World speaker would concentrate on the injustice in their distribution and consumption. Each needs to listen to the other, but often they do not. Most of us, however, saw that it was not adequate to think from the perspective of one's own region or nation, but that all of us had to learn to think globally, and learn from each other's situations, cultural heritages and religious traditions. This is difficult but necessary.

The theological perspectives which dominated were those of Western existentialist or process theologies. Roman Catholic theological approaches were rarely heard. The Eastern Orthodox perspective was often cited in the section documents as a special case, while the Protestant perspective was regarded as the more universal.

Ecumenical dialogue has still a long way to go in this matter of theological ethics, and of learning from other religions or secular perspectives—other than those of the West. There is sufficient evidence that such dialogue can lead to fundamental changes in the perspectives of all participants, though often it takes more time, effort and integrity than we care to invest. There is a certain fear to learn on the part of all who have found their identity and security in a limited perspective. New learning is feared or resisted, where identity is insecure or security of identity is falsely grounded.

We have three major points to make regarding the problem of proceeding from differing theological or ideological starting points towards an ethical consensus: (a) In some cases, consensus is possible even without reference to starting points—e.g., on human dignity, justice in society, or even the unity of the human race. These three values, it seems, have now matured in the modern world as fairly worthy of common acceptance, and should be so accepted. Each religious or ideological perspective should be free to expound and teach these values in its own context, going back to the theological and ideological perspectives in which the values can be rooted. (b) It is important, however, for each of us to understand the *conceptual framework* within which others hold their values, for these are important for the detailed exposition of the values and therefore for their content; it is equally important to sympathetically understand the socioeconomic and cultural-political *situation* from which others speak. This takes a lot of effort and openness. (c) There is real danger in the present pragmatic approach to values, seen sometimes even in Christians. Certain ethical principles, values or preferences may be deceptive. Quite often group self-interest underlies various apparently altruistic value preferences (e.g., aid to developing countries), and we should not shy away from a deeper conceptual framework analysis of all proposed value-systems. One has seen the great psychological tension inside people trained in a rigorously ethical mode of conduct, but whose perception of reality does not see the reason for that ethical conduct demanded either by the trained super-ego or by society. This creates a great rupture between what one is, and what one does. To me this seems to be the root of a great deal of the tension, unrest and psychic breakdown that one sees today in many industrialized societies.

The aversion of our contemporary civilizations to metaphysics and deep philosophical reflection is partly understandable in relation to the Western experience where philosophy has often led people astray. Philosophy can be a source of deception and error and it often has been. But without philosophical-theological vision, people are reduced to a pragmatism wherein also people can be manipulated and misled.

I would therefore enter a strong plea to Christians that they should not abandon fundamental philosophical-theological reflection, and should maintain a constant and active willingness to learn from the traditions of others. I say this with deep gratitude for all I have—be it so imperfectly—learned through the years from so many cultures, European, American, African and Asian. The very diversity of religions and cultures should not be taken in the sense that since no one of them can be exclusively right, therefore all of them must be equally wrong. The transconceptual apprehension of reality lies beyond all human religion and beyond all conceptual grasp; but one moves towards a richer transconceptual apprehension as one passes through a large variety of conceptual and symbolic apprehensions of reality. Plurality is a witness to the many-dimensioned splendor of reality.

THEOLOGICAL ETHICS TO SECULAR ETHICS

The secular language and conceptual framework as it now exists is a contribution of modern science. It has now become a universal language. We can therefore use it in our universities, in our political assemblies, in our international bodies like the UN and its agencies. It is a great instrument of human communication across the globe.

Human beings can now speak to each other about common problems facing humanity, be they Buddhists from Thailand, Marxists from the Soviet Union, Muslims from Saudi Arabia, Protestants from Denmark, Catholics from Argentina or gnostics from North America. This is already a great achievement for which one thanks God at every secular world conference.

Do we now have a language in which to appeal to the conscience of humanity on a global scale? We may think that we do, because we have now a global elite that gathers at world conferences. They do speak a common language. But this common language itself suffers from the limitations of present science. They can speak about resources, skills, personnel, production, distribution, organization, politics, economics, sociology, culture, psychology, physics, chemistry, biology and a host of other disciplines and effectively communicate with each other. They can even talk about values like human dignity and freedom, national and international justice, the unity of humanity and so on.

Yet, it is this very secular language that sometimes inhibits people from speaking effectively to each other about the meaning-structures that underlie our value choices. By meaning structures I mean the implied conceptual or symbolic perception of reality and of goals towards which to advance. Here we have so far operated with two sets of secular systems—that of Marxist humanist ideology or of Western liberal humanist ideology. In both of these, the adjective "humanist" points to the fact that the central concern is humanity and its destiny. The nature of this humanity is conceived in both systems as the highest known product of the evolutionary process. The origin of the process itself is attributed in either system as inherent in "nature" or matter energy in process of evolution. This is an unsatisfactory explanation, since it says that the process is due to the process itself. The process is assumed to be given, self-existent, autonomous.

Those who see the problem of explaining the origin of the process, resort to philosophical dogmas about the meaninglessness of the question since it cannot be answered by science. Even a reputed scientist like Steven Weinberg, who gives the impression in his *The First Three Minutes* that science can explain the origin of the process, resorts to devious ways of reasoning to cover up the fact that the question of the origin of the process of the universe is at present beyond the reach of science, due to limitations in method and framework of conceptualization. The Marxist on the other hand also starts with a quasi-dogmatic assumption: "In the beginning was matter-energy, with the principle of dialectical contradiction contained within it."

Neither can secular scientific thought, whether Western liberal or Marxist dialectical, provide us with much of a clue as to the final destiny of humanity and the universe. Either can fix proximate goals like "a society of happy people", or in the more precise secular language, "the just, participatory and sustainable society", or in a Marxist language, "a classless Communist Society". But these are only proximate goals in history, and do not satisfy the deepest aspirations in man, which include some concern about a personal and common destiny beyond history, beyond death. Nor do they deal with the aspirations of those who want to enter a deeper level of the realization of self—as in the Far Eastern religions of Taoism, Buddhism and Hinduism. And especially because of this failure to respond to the deepest in humanity, our ideological conceptions of proximate corporate destiny in history fail to attract the deepest loyalties—whether it be in Western liberal humanism or in Marxist dialectical humanism.

Failure to deal fundamentally with origin and destiny has direct consequences for the present. Man constantly searches for self-understanding and understanding of the good. These are the basic notions, even when conceptually unclear, which determine the value choices of persons and societies. Precisely because of its failure to deal with origin and destiny, not only does secular ethical reflection become superficial in its value choices; it is unable to draw out the deepest loyalty of man in the pursuit of these values.

The secular movements of modern times are directly rooted in the Renaissance and the Enlightenment, and in the enthronement of conscious and conceptual reason as the source of all enlightenment. Modern science is a child of the Enlightenment or perhaps they are sisters born from a common mother.

But there is no reason why science should be permanently committed to the secular. I believe that the liberation of science from the secular has already begun. I see indicators of this in several current phenomena:

(a)   the frantic search of young people in industrially advanced societies for fulfillment in Eastern religious practices like yoga, meditation and even espousal of Hinduism and Buddhism in various new forms (e.g., Krishna Consciousness, Zen);

(b)   the growing attention by scientists to psychic phenomena hitherto ignored by modern science (e.g., psychical research, more in socialist countries than in the West; scientific studies on bio-feedback, altered states of consciousness, energy and force fields, meditation research, on dying and the mystical consciousness, etc.);

(c)   the growing realization of ecological awareness moving towards an acceptance of the cosmic relations of humanity, i.e., that man is inextricably linked to the whole of reality and does not exist apart from the various fields that constitute the universe;

(d)   the growing centrality of the fable, the legend, the fantastic and the mythological in Soviet entertainment—the most popular ballets and operas today— are those that have something to do with extrahuman entities and forces.

These are all reasons why we should be circumspect about accepting the usual secular ethos of science as somehow indispensable. After all, the very clarity of the concept "secular" is today in serious doubt. Usually it refers to a particular phenomenon of Western history. After a period of domination of all thoughts, symbols, ideas and institutions by the dogmatism and clerical imperialism of the Western Church, these thoughts, symbols, ideas and institutions developed sufficient momentum to break loose from the clerical yoke and sought to establish themselves on autonomous foundations. The very enthronement of conceptual reason was due to the need to fight against the confining yoke of the authority of dogma and tradition. The separation not only between Church and State, but also between Church and socio-cultural institutions like those related to education and medicine as well as art and music, was a direct consequence of the revolt against ecclesiastical domination. The very notion of "secular" in the modern sense was born in that revolt.

It has been a very fruitful revolt and has spawned many ideas and institutions which have now become the common property of many peoples everywhere (hospitals, schools, forms of democratic government). They probably would not have been so universally accepted had they been inseparably attached to the Christian Church and clerical domination.

But does this mean that humanity has to remain forever under the new yoke—the yoke of the secular? After all, the secular is a concept very hard to define or defend philosophically. In essence it meant belonging to the *saeculum*, or time-space world, and not to the Church, which was supposed to be concerned with the open world—that of *eternitas*. Later it was given the definition of time-relatedness, that is, changing with the times or constantly readjusting to temporal change. Still later it was given the meaning that it deals with reality in terms of this world and no other, which meaning was used for an attempt by theologians to interpret the meaning of even the Christian faith in purely this-world terms (Bishop Robinson's *Honest to God*, Paul van Buren's *The Secular Meaning of the Gospel*, Harvey Cox's *The Secular City*, Ted Van Leeuwen's *Christianity in World History*, and the whole Death of God movement in theology, as well as many liberation theologies).

The secular movement is today to be criticized for precisely its unscientific assumptions, for example, that this world can be understood in its own terms without reference to a beginning or source which remains inexplicable in science. The Christian Church should not be afraid to question science's unwarranted imposition of the secular frame on human consciousness. The Church must have confidence enough in its own thought-frame which believes that this universe is not self-existent but created, that this universe is not autonomous but contingent, that this universe has a destiny that is set by God but in the shaping of which human beings are privileged to participate. It is at these points that the Church refuses to be intimidated by the secular tempo of science.

And yet, the secular language remains the only common language in which the peoples of the world can converse with each other and come to common understandings and purposes.

What shall the religious traditions then do? For some traditions like Buddhism and Taoism, which are not conceptually theistic, there seems to be no difficulty in adopting secular language. This, however, is an illusion, for all these religions are heavily dependent on tradition and cannot establish themselves by pure conceptual reason, though many apologists have unsuccessfully attempted such defense in the past.

Tradition thus becomes a key concept in theological or religious ethics in dialogue with secular ethics. Today science itself begins to recognize the role of community and tradition in the maintenance and development of science. True, most of the formulations and theories of science are available in written form; but no scientist acquires all of his skill and knowledge from written books or articles; neither medicine nor surgery as scientific skill and knowledge can be wholly acquired from books. Nor do even the more abstract sciences like mathematics or astronomy become actually transmitted through books alone. There is a constant interaction between teacher and student, between scientist and scientist through which science becomes transmitted and developed in the scientific community. Tradition, of course, generally includes written as well as unwritten elements.

There is a mistaken assumption among some theologians that ethics can be directly derived from the Scriptures, or from the Scriptures plus rational reflection. But when we analyze the process of ethical reflection we find that the ethicist also uses various exegetical traditions in choosing particular passages and in interpreting them. He may cite chapter and verse plus the authority of particular professors. It seems an illusion that one can move from Scripture to ethics without going through tradition.

But the fact remains that one has still to resort to secular language if one is to carry conviction to those outside the particular religious tradition from which the ethicist speaks. This gives rise to what I call "the principle of two languages" in ethical reflection on common social issues. One may arrive at a particular ethical choice starting from tradition, which includes Scripture and its interpretation. Quite often, especially for the Eastern Orthodox theologian (but not only for him), tradition is carried in symbols like the Eucharist, the meaning of which can never be exhaustively conceptualized. The language of tradition thus involves transconceptual experience, and can never be fully translated into secular language. It includes myths, images, and symbols. This is true also in science which is heavily dependent on paradigms and images—e.g., waves, particles, fields, systems, spin, flow and so on.

The task of the Christian theologian is then to have two sets of languages—one of religious tradition, which he uses for discussion among his fellow-religionists and others interested, and another secular language which he uses for conversations with those who accept only the secular mode of reasoning. But the first cannot be exhaustively translated into the second. And quite often when new questions arise in the second, the theologian has to go back to the first in order to derive inspiration and guidance.

Even more significant is the fact that neither of the languages is static. The theological traditions, when it refuses to learn from the new insights of science and secular thought, becomes stale and irrelevant. A constant interaction between the two languages and the two modes of awareness seems essential for the vitality of both traditions. Science, too, needs to develop the two-language system, especially as each discipline becomes increasingly more complex, technical and incomprehensible to outsiders. There will be an inside language, a sort of shorthand which the disciplines use internally; but then they must also learn to speak to those outside in a more general and less technical language.

Only as both theologians and scientists develop an adequate proficiency in such two-language structures can we hope to have communication and genuine interaction within the community. This does not mean that outsiders must be denied access to the technical language: it should not be used by any group to prevent communication with the outside. The function of language is not to obstruct communication but to facilitate it.

There must be frequent passage between the two languages, both in theology and in the scientific disciplines.

ADVOCACY ETHICS *VERSUS* FREE ENQUIRY IN ETHICAL REFLECTION

Here, too, we face a number of problems. Are there certain ethical values to which we cling without compromise, while in other areas the churches' role may be to promote and facilitate somewhat free inquiry? It seems the ethical values to which we can cling are of a general nature —love, joy, peace, justice, human dignity, the unity and equality of worth of all human beings, etc. But there are so many other values which are furiously disputed in the churches—e.g., nonviolence, reverence for all life, the conflict of rights between mother and embryo, the right to live, the right to die, etc.

Clinging to a value like love or peace does not necessarily mean that one does not fight against injustice and evil. But should not the value of love or peace undergird even the fight against evil? That is to say, adhering to a general value like love does not mean that there cannot be free discussion about what that value means in a particular instance or how it comes into conflict with other values.

People expect the churches to *stand* for certain values, not merely in the general sense of love, joy and peace, but in more specific terms like a clear position for or against abortion; for or against nuclear energy; for or against war. Many Christians want the churches to tell them what they should or should not do. This temptation to depend on authority itself needs to be carefully examined in ethical terms. The word of the Church cannot be yes and no *at the same time*. But neither can it always be either yes or no.

The Eastern Orthodox Church did develop a system of legalistic ethics at one time. Basil the Great (ca. 329-379) witnesses to an ancient tradition in the Church which forbade abortion and prescribed ten years of penance to a woman who procures abortion.

Nevertheless the Eastern Church puts its major emphasis not on sin and punishment, but rather on separation from evil and growth in the good. On an issue like abortion, the Eastern Church takes a firm stand against it. But on most issues, the Church does not legislate. Its task is to help people to overcome their proclivity to evil by discipline and, where necessary, by punishment. However, it seems more important to emphasize the overcoming of the desire to do evil, and not merely to avoid overt acts of evil-doing.

In Christian ethics, too, the Eastern Orthodox would therefore emphasize the positive aspect of growing in the good, so that even if there were no law against a particular evil, one would not want to do it. The emphasis falls on growth in the good, in the capacity to discern what is good and what is evil, and being the kind of person who finds fulfillment in the doing of good—the accent is on *being* good rather than *doing* good, which should spontaneously flow from a good being.

In that context, while on some clear and specific issues like murder, abortion, adultery, stealing, etc., the Church takes a clear advocacy role, on most issues the important thing is to promote free enquiry which helps a person to rightly discern between good and evil, with full awareness of why something is evil or good.

One could thus say that ethics, which deals primarily with external actions, is not sufficient. One has to activate the power of the Holy Spirit in the community of faith in order that persons may grow in the discernment of good and evil, and grow in being which separates itself from evil in mind and will and advances in the love of the good. Heroic acts of good are more important than the avoidance of evil, though the two have to go together.

Free enquiry and reflection on moral issues is thus not to culminate in moral legislation, but in helping people to develop their powers of discerning between good and evil and choosing the good by their own free will rather than by legislative compulsion.

CHAPTER EIGHT

Science and Faith

Towards a New Partnership

TOWARDS A UNIVERSAL CHRISTIAN HUMANISM

In 1934 the famous Roman Catholic Professor Jacques Maritain delivered six lectures at the University of Santander on Integral Humanism.[1] Today this work (published in English in 1968) stands as an almost singular recent instance of an integral Christian approach to the philosophy of human action. Nothing comparable has been attempted by Protestant or Orthodox theologians.

We are today in need of a fresh and ecumenical approach to such an integral Christian vision of reality and of our task in it. Here the best of science and the best of philosophy must integrate itself with the best of ecumenical theology to provide a coherent, provisional, dynamic vision.

Such a philosophical approach by necessity has to be speculative, as Maritain says:

Practical philosophy remains philosophy, it remains a knowledge speculative in mode; but unlike metaphysics and the philosophy of nature, it is ordered from the very beginning to an object which is action, and however great may be in it the role of verification of fact, whatever account it must take of historical conditionings and necessities, it is above all a science of freedom.[2]

It is this science of freedom that we need to develop in our time, not merely Christian social ethics in the old categories of context and principle and decision-making. Here science and faith have to enter into a new partnership with true philosophy—i.e., the love of wisdom.

To put it in other words, humanism needs a new Christian basis, which takes into account not only science and philosophy as they have developed in the modern West, but also the wide range of other religions and cultures. It cannot be based on a mere critical liberalism—which avows mainly a general scepticism, a refusal to accept the authority of tradition, an aversion to dogma and creed and a trust in the ability of reason and good sense to solve all our problems.

Such a Christian humanism cannot be the blueprint for a New Christendom which simply integrates into the Old Christendom the now aging Western Liberal Humanism or even Western Socialist Humanism based on the metaphysics and philosophy of history of Marx and Engels. We must learn from all these, but we must overcome the vain hope of a Christian imperialism, whether in ideas or in action.

Jacques Maritain proposes his *integral humanism* "which would represent for them (i.e., Christians) a new Christendom no longer sacral but secular or lay. . .which has no standards in common with 'bourgeois' humanism because it does not worship man but really and effectively respects human dignity and does justice to the integral demands of the person as oriented towards socio-temporal realization of the Gospel's concern for human beings . . . and toward the ideal of a fraternal community."[3]

Jacques Maritain brings to bear upon his vision of the

future the best in the Western Judaeo-Christian tradition. Brought up as a liberal Protestant, married to a Russian Jewish intellectual, converted to Roman Catholicism, this outstanding modern student of St. Thomas Aquinas is without peer in proving a Western Judaeo-Christian synthetic vision for the future of humanity.

But even for many Western thinkers outside the Roman Catholic fold it does not provide an adequate frame for striking up a new partnership between Science and Faith; for Maritain regarded a worldwide network of Roman Catholics engaged in a new kind of "Catholic action" as the means of redeeming the future.

We can here only seek to outline certain orientations for a new partnership between Faith, Science and Philosophy where all are at the service of humanity, but in doing so we need to go beyond the European faith, European science and European philosophy, which Maritain has not managed to do.

Such a universal Christian humanism must necessarily learn from other religions and faiths and from other philosophies, and should provide new orientations for science other than the domination of "nature". And it must be rooted in the faith of the Church. It must, however, be formulated in such a way that no religion, including the Christian Church, would hold a privileged position in such a humanist society. It will permit various religious and nonreligious perspectives to flourish side by side, but will not seek to impose the secular perspective by driving religions out of the universities and other institutions of society.

Such a humanism will need a dual statement—one that is directly related to Christian convictions and symbols and beliefs; the other stated in nonreligious terms so that it can be acceptable to adherents of other religions or of none. The religious statement must be there; it must be public; it should be open to criticism by Christians and non-Christians alike, but its presence is a necessary safeguard to prevent the domination of an exclusively secular perspective which can always be enslaving.

TOWARDS A MORE UNIVERSAL UNDERSTANDING OF FAITH

Faith, as well as Science, has its own paradigmatic assumptions. No analysis of faith can be independent of the paradigm of reality held by the believer. If, for example, you understand God, Man and World as three separate and disjuncted realities, then you will have a particular conception of faith. If, on the other hand, you hold the view that Man and Universe are integral to each other and that the two together exist *in* God, then your view of faith as well as of science will be different.

Section II of the Conference, which dealt with the topic "Humanity, Nature and God", was unable to enter into the depths of this problem. Section I, which dealt with the Nature of Science and the Nature of Faith, seems to have been biased in favor of the disjunct view of God, humanity and universe. Christian faith was defined as follows:

> Christian faith involves an activity and receptivity of the self in relation to God, which it expresses in the value-laden images of the faithful community of the Church, whose origin and continuance are regarded as the work of God, and whose historical focus is the life, death and resurrection of Jesus Christ.

The self acts and receives, as if standing outside God, from a God who is outside of oneself. This conceptual framework would naturally emphasize the "encounter" aspect of faith—a personal encounter between one's person and the person of Christ, leading to repentance, faith and obedience to God.

In the nondisjunct paradigm, faith is seen as the activity of the Holy Spirit who removes the alienation between Man and God brought about by sin. The consequence is an experience which can be conceived, not as an encounter with someone outside of one's own self, but as a realization of one's true being as rooted and grounded in the Person of Jesus Christ and therefore in the Holy Trinity itself. In the same moment of realization one sees also that the rest of the universe does not exist apart from or outside of God. To be in the Holy Trinity is the only true possibility for the universe to be; for to be "outside" the Holy Trinity is impossible for any being, since there is no "outside" for a God who is in-finite, that is, without boundary. To be "outside" would thus mean "not to be". Insofar as the universe participates in being, it has to be within the creative *energeia* of God; that is the only place for the self also "to be". Thus Man and Universe, bound together in the same contingency of existence, find their true being only *in* God and not over against Him.

If faith thus means the new reality of experiencing one's own and the universe's rootedness and groundedness in Christ, a new perspective follows also on the activity of science as something that takes place "within God", a new luminescence and a new interaction within the package Man-Universe, grounded in and contingent upon God's Trinitarian dynamic being.

But there is an intrinsic difference between the Creator God's Being and the created being of humanity and universe. This difference is expressed by the ancient patristic philosophy in the following terms:

He Who Is	**The Things That Are**
(Being of God)	(Existence of Creation)
(a) derived from itself; not owing its being to any other entity or dependent upon anything else; good in itself.	(a) has its source outside itself and is contingent upon that source; apart from that source it has no being or existence; being and good only by participation.

(b) The is-ness of God is beyond human comprehension; there are no conceptual categories with which to grasp the mode of God's being —neither analogy nor image.

(b) The existence of persons and beings in creation can be grasped at least in part conceptually—in terms of their time-space location, their species and genera, their origin and function, relation and purpose, their intention, etc.

(c) God wills what He is, and is what He wills. He thus has no need to *become* something else than what He *is* (this is quite contrary to Process Philosophy which regards it necessary for God to realize his potential being through becoming in time).

(c) In creation, all things are in the process of becoming, or going out of existence. Nothing remains unchanging. All beings come to be, become, and then either continue to become and grow, or begin to go out of existence. No created being is free from the need to become.

(d) Neither temporal nor spatial extension belongs to the Being of God—hence no spatial distance between the Three Persons of that Being, nor temporal before and after.

(d) All things in creation are extended in time and space though there may be different kinds of, or experiences of, time and space.

Science is an activity taking place within God, not outside of Him. But being "inside" God does not necessarily mean "in the bosom of God", or in the actual incomprehensible *ousia* (being) of God, but within that realm within God where His energies operate. Science does discover some of the regularities and predictabilities within that operation, under certain conditions.

Potentially, science has the capacity to explore many areas within the operative energies of God, though so far it has only touched the fringes—a little knowledge of how "nature" operates on the macro-level, with less knowledge of the micro-level; including various aspects of physics, mathematics, astronomy, chemistry, biology, psychology, sociology, economics, politics, etc. Such *regularities* and operationally useful predictabilities as science uncovers belong to the realm of God's operative energies, but so does the *knowledge* of them by man. It is God's *energeia* that is studied; it is the same *energeia* in man which studies it.

The problem with science, however, remains crucial, i.e., though it takes place within the *energeia* of God, yet, so far as the subject-object dichotomy remains, there is an element of alienation in the knowledge produced by science—the knower, the known, and the knowledge remain somehow slightly disjunct and integrated only in an external way.

In faith, as understood by the Eastern Orthodox, there is an experience of a partial overcoming of this alienation. The true Christian believer experiences the uplifting, supporting, nourishing presence of God, not as something over against him, but as something on which he is established as on a rock.

Here the doctrine of the Holy Spirit is central to our understanding of faith itself. Equally important is the doctrine of the Church. It is in the divine-human community of the Body of Christ that the Holy Spirit provides the experience of disalienation which is called faith.

For the Eastern Orthodox, faith is not an action of the self, but a divine-human action in the community of faith, an action in which the Holy Spirit makes it possible for human beings to be *incorporated* into the Body of Christ and thus to *participate* in the inseparably united divinity-humanity of Christ. It is not my personal act of a "leap into the unknown", nor is it my "believing against the understanding", nor my "subjective determination", as Kierkegaard in much too individualistic and anthropogenic a fashion, defined it. Faith may sometimes give rise to a "passionate inwardness" holding fast to "an objective uncertainty with the passion of the infinite".[4] But faith cannot be defined in those terms.

When a ten-week-old child is incorporated into the Church by Baptism, the child's "passionate inwardness" is related more to the temperature of the baptismal water than to any holding fast to an objective uncertainty. It is the Holy Spirit, operating through the faith and action of the community, that introduces the infant into the Body of Christ, there to participate in Christ, through Chrismation and the Eucharistic communion which immediately follow upon baptism.

Faith, too, is thus an activity of the Holy Spirit through the community of faith, though each person is free to grow in faith or to grow out of it. Neither faith nor science are primarily individual activities. They both take place in communities: the Holy Spirit of God is present in both communities though in different modes and operations.

TOWARDS A WIDER AND DEEPER UNDERSTANDING
OF THE HOLY SPIRIT IN RELATION TO FAITH AND SCIENCE

Since our apprehension of the Holy Spirit is very limited, and since the Spirit by nature does not draw much attention to itself, we can never draw an adequate conceptual apprehension of the Third Person of the Trinity. The best we can do is to make ourselves aware of some of the Spirit's *operations*. The *Being* of the Holy Spirit remains incomprehensible, since it is the One Being of the Triune God.

Among the operations of God the Spirit, we can distinguish between two realms, traditionally distinguished as Creation and Redemption. It is best to see that operation as the creative *energeia* of God, but within that single operation which is the created order, we can distinguish two specificities—the *general* operation which brings the creation into being out of nothing and leads it to fulfillment, and the ecclesial operation which was initiated in the incarnation of the Lord Jesus Christ and in the special dispensation of Pentecost, and which continues in the community of faith. The two do not exist in two separate spheres: the incarnation occurs within the Created Order and is a new stage in the overall operation which brings the creation to its fulfillment. It is only for the sake of greater clarity that we make the distinction between the general operation and the ecclesial operation.

The danger in Christian thought is to confine the Spirit's work to the specifically "religious" operation within the Christian Church, or in the inspiration of the Christian Scriptures, or in revelation, or in the individual consciousness of the believer. We can get certain glimpses of the general operation of the Spirit in the Old Testament. But it will be unwise to limit ourselves to a scientific exegesis of the Old Testament to understand that general operation. What we have in the Old Testament are merely certain pointers to that general operation.

The New Bible Dictionary, a conservative English reference book put out by the Intervarsity Fellowship, enumerates five different aspects of the Work of the Spirit in the Created Order:

(a) The Spirit, brooding over the primeval waters (Gen. 1:2), and creating humanity (Gen. 2:7), sustains animal and plant life (Ps. 104:30) and gives humanity its whole psychic and physical powers.

(b) The Spirit as the equipper for service by giving individual men special skills and powers (Ex. 31:3, Judges 3:10, 14:6, etc.).

(c) The Spirit who inspires, operates and speaks through the prophets (e.g., Isaiah 63:10, 11).

(d)   The Spirit as creator of humility, repentance, a clean heart, constancy and joy in people (the Psalms, esp. 51, 139, etc.)

(e)   The Spirit as foretelling the coming of the Messiah (Is. 51:2-9, etc.).

In this work as well as in the modern Roman Catholic *Sacramentum Verbi*[5], the Spirit of God is interpreted in too narrow a "religious" setting, seeing it as "that mysterious force which proceeds from God and takes powerful effect in the history of the covenant people" (*Sacramentum Verbi*), seeing the prophets as the "bearers of the spirit of God *par excellence*".

We need a wider understanding of the operation of the Holy Spirit, enriched by our insights from the Old and New Testaments, but going beyond to learn the new things which the Spirit has taught about its operations in the whole tradition of Christianity. As St. Gregory Nazianzen put it:

> The Old Testament proclaimed the Father openly and the Son more obscurely. The New manifested the Son, and suggested the deity of the Spirit. Now the Spirit Himself dwells among us, and supplies us with a clear demonstration of Himself.[6]

And further along in the same work, Gregory waxes eloquent about

> The Spirit of Wisdom, of understanding, of Counsel, of power, of knowledge, of Godliness, of the Fear of God. For He is the maker of all these, completing all with his being, holding all things together, fulfilling the cosmos in accordance with its being, yet incomprehensible to the world in terms of its dynamic power, good, straight-forward, Lord by nature and not by commission; sanctifier, measurer not measurable; participated in, but not participating in; filler of all not needing to be filled; containing (all) but not containable; inherited (by us); glorified; connumerated (with the Father and the Son); the subject of serious warning (not to sin against); the Finger of God; Fire as God, to emphasize its consubstantiality with God it seems; the Spirit who made (all things)

and creates anew by Baptism and the Resurrection; the Spirit who is knower of all things, the teacher, the Wind who blows where it wishes and as much as it wants to; the Guide, the Speaker, the one who commissions and marks out boundaries; the wrathful whom people tempt; the Unveiler, Illuminator, Life-giver, rather is itself Light and Life, the Edifier of the Temple of God, the Deifier. . . .[7]

This is how the ancient fathers of the Church understood the Holy Spirit. It is the source of all knowledge and wisdom, all skill and power. Why should we place the human activity of science and technology as having a source outside the Holy Spirit? Of course science and technology can become demonic, just as faith can become demonic. But the source of all true knowledge and skill is the Holy Spirit, whether in the Created order in general or in the Church. All the gifts of the Spirit are however given in freedom and can be misused for destruction of oneself and others.

Once we recognize that science as well as faith come from God by the power of the Holy Spirit, we are on our way to a properly Christian integration of Science and Faith. We can make distinctions between the operation of the Holy Spirit in the created order and that in the Church; the operations can be differentiated, but the source is One.

In physics or politics, in economics or in biology, in the world or in the Church, all genuine and true illuminations and clarification comes from the Spirit. The ruler and the law-giver, the bishop and the scientist, the computer technologist and the spiritual counselor, all get the right skill and knowledge from God the Holy Spirit. Art and science, philosophy and faith—all are from the operation of the Spirit.

Science, faith and sin

One may find the idea that the power of science and technology is part of the work of the Holy Spirit within the created order a bit too optimistic and uncritical. The work of the Holy Spirit has to be seen, however, in the context of a greater apprehension of human freedom and sin, as well as of the Eastern Orthodox understanding of the principle of *synergeia*.

The concept of sin cannot be understood apart from the concept of freedom in the created order. And in understanding science as well as faith, freedom is the key category. Since we have dealt extensively with this topic in other works[8] we will be brief here in recapitulating the main points.

(a)   Freedom is to be seen in both aspects, i.e., freedom *from* and freedom *for*. Freedom *from* refers to liberation from external and internal constraint that prevents freedom *for*. But the removal of the constraints does not automatically generate freedom *for*. The latter depends on *creative power*, not only to choose something, but also to realize that which is chosen. Freedom *from* sin does not necessarily bring freedom *for* righteousness. All genuine freedom involves both movements—the freedom *from* external and internal constraints and the freedom *for* creating the good. Both these aspects of freedom have to be *won*, both as a gift of grace and as a consequence of disciplined struggle.

(b)   God alone is truly free in both aspects. He is not only free from all internal and external constraints, but has infinite power and wisdom to create what He chooses. In fact, God is so free that His will and Word, which are always coincident, immediately become reality. The created order is the manifestation of God's freedom. It is His will and Word, in the infinite freedom of creative power, that has given birth to the creation and sustains it in existence today.

(c)   When God creates humanity in His own image,[9] the image becomes endowed with the same freedom; though not in the same infinite manner like the original. Humanity was originally, as created, free except in one external constraint—not to taste the fruit of the

Tree of the Knowledge of Good and Evil. But the restraint was only in the form of a command; Adam and Eve were not physically restrained from eating the forbidden fruit; they were *free* to do so at the price of disobedience; and they exercised that freedom to disobey and thereby lost most of the positive freedom they had—to create the good. Most of it—because they could still love and care for each other and create some limited good, though the evil of saying no to God, which they had created in their freedom, encroached upon even the good they did. From them come both Cain and Abel, both evil and good. They are, however, prisoners of evil and therefore of death; even the limited good they create is soon invaded by evil.

(d)   Sin, the act of freedom, becomes an alien power that controls humanity, and enslaves humanity to the three-fold master: sin-law-death (Romans 5:8). But Christ frees humanity from the enslavement; humanity is now free to overcome sin, law and death and to live in the freedom of creative good—by the power of the Holy Spirit working in and through humanity. Sin, however, continues to be active in the believer and in the unbeliever, and in the social structures in which both live together.

(e)   Sin creates alienation at various levels—between the human self and God, between the human self and other human selves, between the human self and the structures of social living, between the humanity and the rest of creation, and even between the human self and its own existence. This breaking-up of relations, this distance and lack of communication between existents, and this sense of threat or anxiety about the other encroaching upon one's territory—all these are manifestations of sin, or consequences of a fundamental rupture between existents and the source of their being. Sin creates the possibility of total fragmentation, and initiates a process of dissolution which ends up in nonbeing. But even the fallen creation has not lost all its links to the source of its being, for if it had, it would have been instantly reduced to nothing. History is a process when sin and righteousness coexist, and human beings even in their sin hunger and thirst for righteousness. And even in generally decadent societies, occasional lamps of righteousness are lit, and heroes of the Spirit arise from time to time, keeping aloft in some form the ideal of righteousness.

(f)   But since the Incarnation, Death and Resurrection of Jesus Christ, there is a totally new status for the created order, and especially for the fallen part of creation. God is now personally present in his very being in this fallen creation; the fallen creation is now in a new situation of freedom—to live in the new creation initiated by the incarnation of the Son of God, by the powers of the Spirit present in the fallen and redeemed creation, or to continue to live in the old decadent order and be subject to the powers of discord, death and dissolution.

The faith community as well as the science community exists in this new situation of freedom. Every choice for the good is a choice for the new; every act in the new situation of freedom, whether it be in the Church or in the world, inevitably involves such choice, either for the life-giving new or for the self-destroying old.

Faith is the way to a conscious participation in the new; but even those who do not profess the Christian faith do participate in the new by virtue of their choices for the good. For the new is not by any means limited to the Church.

What advantage then does faith give? One can enumerate a few in a very brief way:

(a)   Faith delivers persons from all fear of the future and worry about past guilt, from fear of death and anxiety about condemnation, establishing the person on the firm foundation of Christ and opening up channels to the powers of God available in the new.

(b)   Faith provides confidence that the future of all is safe in God's hands, that evil cannot finally triumph and that the good will be finally liberated from the mixture with evil. This gives one the courage to face the power of evil, to challenge it, and where necessary to accept martyrdom.

(c)   Faith gives deeper insights into the ways of God's working in the Universe and makes it easier to work with and not against the purposes of God.

(d)   Faith initiates one into the community where all the means exist for one's being edified or built up as a member of the Body of Christ, through a progressive separation from evil and growth in the good.

(e)   Faith provides persons and communities with the courage for integrity and self-sacrificing love, since the knowledge of the grace of God in Christ frees one from the need either to justify oneself or to seek one's own. This integrity and love manifest themselves in new ways of beneficent creativity.

The fact that many of these advantages and possibilities are not always appropriated and realized by persons in the community of faith points to the phenomenon of sin which invades also the community of faith and persons participating in it.

Faith, as well as Science, should provide the possibility for people to exercise their freedom in the fight against evil and in the creation of the good. The presence of sin in the structures of the fallen creation makes both faith and science vulnerable. Sin is allowed freedom to militate against our true freedom. This is the tragedy of both faith and science—the tragedy of failure to exercise rightly the new freedom given in Christ for overcoming evil with the good. Science-technology is a new God-given arena of freedom, where new diabolical possibilities of evil and heroic possibilities of good coexist.

The new partnership between faith and science has

thus to be based on recognizing their common source in the operations of the *energeia* of God, and their common vulnerability to the power of nonbeing or sin or evil, which continues to operate in history. Faith then no longer wants to control science; nor does science claim to set limits for faith. The two learn from each other, correct each other, and respect each other. They also acknowledge their own limits, neither triumphantly claiming access to all knowledge and truth, nor wringing their hands in abject despair about human sin and fallibility. Within those limits, both science and faith can cooperate with other human endeavors like art and philosophy, music and literature, love and mercy, efforts for peace and justice and so on, to show the way for shaping a world and a humanity that more faithfully reflects the glory of God, which is also the glory of Man.

This is the new challenge for both faith and science—not merely to coeixst in an uneasy truce, but genuinely to collaborate in creating greater visions of the good and working together to realize these visions.

CHAPTER NINE

New Orientations in Faith and Science

Faith has often been too narrow-minded; the biggest challenges before the Christian faith community today seem to be the following:

(a)   to overcome its cultural parochialism, by which it makes its expression in a particular culture, time and place universally normative;

(b)   to overcome its tendency to totalitarianism, restricting the freedom and liberty of people to think or act differently;

(c)   to overcome its preoccupation with the salvation of individuals alone, and become concerned both with the building up of persons in the Body of Christ, as also with the rest of humanity and in fact with the whole creation;

(d)   to open itself up to learning from other cultures, religions and ideologies, when necessary revising its

own paradigmatic framework and understanding of
reality;

(e)   to reinforce its true being with a better balancing
of the symbolic-cultic, practical-ethical and intellectual-
ideological expressions of the Christian faith in the
light of its apprehension of how God's reality operates;

(f)   to recognize the corrosive presence of sin within
the faith community and continually to manifest repentance, self-criticism and a desire to make amends
irrespective of the cost;

(g)   to discipline itself to be more open to the healing,
correcting, illuminating and creative powers of the
Holy Spirit.

In order to do all this, the Christian community of
faith will have to radically rethink some of the pernicious
dichotomies which have plagued its thinking in the past,
especially since Augustine of Hippo made some of these
dichotomies somewhat respectable in the West.

(a)   The false duality between the city of God and the
city of the earth as two mutually opposed realities has
perhaps done the greatest damage. The idea was that one
has to take one's heart away from its love for the city of
the earth in order to love the City of God. A good eschatological perspective should be able to see that elements
of the City (Kingdom) of God can manifest themselves,
though imperfectly, in the City of the Earth. The need is
thus not to pluck your love away from the city of the
Earth, but to love it in such a way that more and more
elements of the city of God become manifest in the City
of the Earth.

(b)   The false dualism of matter and spirit may be of
Indo-hellenic origin, but its equation with the evil-good
dualism has done havoc to Christian thought. If Hegel saw
Spirit as the Absolute, Max and Engels saw Matter as the
primordial Absolute. Today the Marxists as well as Christians have begun to see that matter and energy are inter-

changeable entities and that what we call spirit is nothing but matter-energy in a more evolved form. Matter is not an enemy of the spirit, but its less evolved form, its vehicle and form of manifestation to the senses, its instrument and medium. The Incarnation of Jesus Christ in a material body and the translation of that material body into the heavenly[1] realm through the Resurrection and Ascension of Christ should have taught us not to despise matter. Even the doctrines of the bodily resurrection and of the Eucharist did not help us to overcome our deeply ingrained Gnostic neo-Platonic distrust of matter. Even today we hypocritically curse and rail about materialism, as if matter had not been created by God and were somehow alien to Him.

(c)   We have already overstressed word and concept, to the detriment of symbol and ritual. Humanity cannot grasp the transcendent truth in word and concept; it must give expression to its deepest perceptions through rituals, sacraments, community liturgies, through dance and music, painting and sculpture, architecture and literature, myth and legend. We have to overcome our over-cerebration by becoming more celebrative. Theology has too often claimed to capture the truth. The word and sermon have been polluted by over-use, and must regain their integrity through a period of more disciplined silence and more expressive action. To be exclusively word- and concept-oriented is a male middleclass sickness, which has to be overcome by a balancing of storytelling, music and nonverbal expressions.

(d)   Theology must overcome the false duality between "vertical" and "horizontal"—the assumption that there are some things which involve a one-to-one relation (vertical) with the God above, and others (horizontal) which involve relations to fellow human beings without thereby involving God. If a proper nondisjunctive paradigm of God, humanity and universe is assimilated as the basic framework, then several false dualities can be over overcome— vertical-horizontal, nature-grace, natural-supernatural, matter-spirit and so on. One's prayer (vertical) then becomes saturated with socioeconomic concerns and the needs and interests of others (horizontal). There will not then remain one realm (nature) where man is master and another (grace-supernatural) saving and healing activity as coming from God, and the sacred-secular or sacred-profane distinction itself would lose its importance.

(e)   This of course would mean the development of a doctrine of sin which deals with all aspects of alienation in the created order, not just with so-called personal sin or violation of some preconceived moral code. Sin affects person, society and cosmos; redemption in Christ by the Spirit must also affect all three realms. This kind of Christology-Pneumatology and the resulting ecclesiology still need to be worked out. If sin is personal, social and cosmic alienation, then salvation must mean disalienation or reconciliation in all three dimensions.

(f)   This would also mean that our general notion that faith receives its challenges from naturalism, materialism and secularism will also have to be rethought. These world views exist mainly because the Church in its institutional manifestations has failed to give an adequate basis for life—in its worship, practice and thought. We do not need to fight naturalism, materialism, and secularism, but rather to correct our own misstatements and malpractices which have led to the development of these systems. Naturalism, for example, is an offshoot of the kind of Deism that the Church propagated at one time. Materialism, so-called, is a reaction against the kind of ethereal spiritualism we preached. And secularism is a revolt against the arrogance of the Church in seeking to control and dominate all forms of human self-expression—science and art, ethics and philosophy, institutions and processes. A more free and honest approach to other people's ways of thinking, acting and worshipping will help purify society to a great extent through honest self-criticism by Christians as well as others.

Christianity, if it does any fighting at all, should wage war against deeply entrenched institutional, intellectual and spiritual sin in its own bosom. This war against its own alienation will be concurrent with a similar war against deeply entrenched institutional, intellectual and spiritual sin in society.

But faith fails when its criticism of either the Church or of society does not spring from love. Too often it is the desire of a small group to justify themselves or to feel superior which becomes the source of our criticism. All verbal criticism thus becomes spurious, when one's compassion and love for all fail to find some place in that criticism.

Faith receives its challenge from a recognition of its own failures, rather than from a mythical entity called the "modern mind". The mind of man will not necessarily always acknowledge or accept the truth, but sooner or later it seldom fails to recognize genuine and authentic love.

NEW ORIENTATIONS FOR SCIENCE—PHILOSOPHY OF SCIENCE

The present author, not being a practicing scientist, is hardly qualified to say anything worthwhile about possible new orientations for science. Yet, as an amateur in both the human race and in science, one dares to state a few concerns.

One lesson which the author has learned from his limited contacts with many outstanding scientists is that most scientists have not had, as part of their training, much work on the nature of the scientific enterprise either from a sociological or from a philosophical perspective.

If scientists, as part of their training, were to get some grounding in the philosophy and sociology of science, they might also get a better understanding of what in fact they were doing and how it fits into a society characterized by injustice and alienation. There are too many myths about science prevalent in society, e.g., that science provides objective, proved knowledge about all reality, that it is value-neutral, that the problems are connected only with how science is used and not with the nature of science itself and so on.

We need a generation of outstanding scientists who are also philosophers and sociologists of science at the same time. This is necessary for a reorientation of scientific research by scientists themselves who have understood something about how science/technology affects human existence and the human predicament.

Deepening the studies of the philosophy of science can lead to a large number of fruitful insights about our predicament. At present we have mainly three or four noticeably distinct trends of development in the philosophy of science:

1. *The English-Speaking School:* The Vienna circle of discussions (*Wienerkreis*), having landed in the English-speaking world during the rise of Nazism, gave birth to quite a crop of philosophical approaches ranging from logical positivism to linguistic analysis. Again, following a high infant mortality rate, the surviving schools of Empirical philosophy are seeking to come to a consensus, though debates like that between Popperians (following Sir Karl Popper) and Kuhnians (a lesser tribe following Thomas Kuhn) go on still. Some of the fundamental questions posed in the debate bear witness to some basic ambiguities in scientific knowledge. David Hume had already posed it in the eighteenth century: "Are we justified in reasoning from (repeated) instances of which we have experience to other instances (conclusions) of which we have no experience?"[2] Hume said no. But if that answer were right, Bertrand Russell's conclusion would logically follow: "Every attempt to arrive at general scientific laws from particular observations is fallacious and Hume's scepticism is inescapable for an Empiricist."[3] Popper tried to qualify that brutal conclusion by suggesting that Hume meant something else—namely that people do have many nonrational ways of arriving at conclusions and though scientific theories are not strictly rational, they are still useful for survival and therefore justified.

British empiricism has now taken on a more modest pose about the validity or truth-value of the conclusions of science. The position once held by Gilbert Ryle that scientific laws are established and not conjectural is no longer acceptable. Popper, for example, would contend that all scientific laws and theories are conjectural, but some conjectures are preferable to others, because they yield better results and stand up much better to logical refutation. "The method of science is the method of bold conjectures and ingenious and severe attempts to refute them."[4] Popper's understanding of science is largely Darwinian, to the effect that human conjectures struggle with each other for survival, and that only the fittest or best adapted do survive; the mere fact of having survived is evidence enough that it is closer to the truth than the ones that perished.

Scientific enquiry, according to Popper, does not begin, despite popular myths, with experience or observation. It has its beginning in problems, which then find solutions which are found to be inadequate; then follow struggles among various solutions, refutations, new conjectures—a whole bloody mess just as in natural selection; finally one species emerges and still has to struggle with newer and better solutions.

> Problems of explanation are solved by proposing explanatory theories; and an explanatory theory can be criticized by showing that it is either inconsistent in itself or incompatible with the facts, or incompatible with some other knowledge.[5]

This understanding of science is, of course, based on the three-fold theory of the truth-value of propositions— the truth of a proposition being judged by its correspondence with facts (correspondence theory of truth), its coherence within itself and with other beliefs and convictions and experiences (coherence theory of truth) and what is not quite clearly stated but implied in Popper's definitions, the practical or operational value of a proposition (pragmatic theory of truth).

Popper would claim that the first (correspondence with facts) is the primary test of the truth-value of a proposition. But of course propositions and facts can correspond with each other only in a meta-language which has commensurable denotations for facts and propositions. This is the sense of truth held by Alfred Tarski. If I want to speak about statement S and fact F I must have a language which can speak about both Ss and Fs.[6] And truth is an equivalence between S and F.

Thomas Kuhn in his *Structure of Scientific Revolutions*, thought this was too simplistic and wanted a basic distinction made between "normal science"—everyday scientific research firmly based on previous scientific achievements acknowledged by a particular scientific community, and a "scientific revolution" which fundamentally alters the basic paradigm or framework within which scientific understanding takes place. At one time Newton's *Principia* and *Opticks*, Franklin's *Electricity*, or Lavoisier's *Chemistry* provided such an accepted paradigm. Within the Newtonian mechanistic paradigm of reality, where everything is matter in motion according to the laws of mechanics, light was seen as composed of material corpuscles. Then somebody comes along in the early nineteenth century (Young and Fresnel) to suggest a new paradigm for the understanding of light—as a transverse wave motion. The corpuscular and undulatory paradigms compete, each being capable only of a partial explanation of the observed phenomena, until a revolution takes place with Max Planck, Albert Einstein and others—the quantummechanical paradigm which proposes the notion of photons or corpuscular entities which are undulatory. The change from one paradigm to another does not take place in normal science; paradigm change is scientific revolution. Science, according to Kuhn, does not progress as much by Darwinian evolution as by these revolutionary jumps which are quite frequent, and which, unlike normal science, bring substantial amounts of new information.

The debate between Popper and Kuhn in the English-speaking world has proved to be productive of more heat than light. Paul Feyerabend contributes his mite by arguing *Against Method*;[7] he does not believe in law and order science, but advocates an anarchistic theory of knowledge. The imposing of methodological rules and regulations, Feyerabend claims, would stifle the creativity of science.

Much of great science was achieved by violating the rules. If practicing scientists listened to the philosophers of science, they would undergo the fate of Galileo and there would be no more progress in science.

All these and a few other divergent views about what actually happens in the scientific enterprise were put together in a British University Symposium and the results published.[8] Professor Lakatos, formerly of London University, one of the editors of the Symposium report, has made certain observations in his paper, which are yet to attract the attention of intelligent people:

> Now very few philosophers or scientists still think that scientific knowledge is, or can be, proven knowledge . . . But few realize that with this the whole classical structure of intellectual values falls in ruins and has to be replaced.[9]

Scientific theories are operational statements; they can be neither proved nor disproved. They can be rejected, when the community decides that a "better" one is available which has more predictive and explanatory power and is more elegant. There are some general indications on how to decide whether one theory is better than the other; but no rules can be laid down for making such decisions. The consensus in the community is decisive for the rejection of one theory and the acceptance of another.

So finally the consensus among philosophers of science in the English-speaking world is that no scientific theory is final; it is the best so far, until something better arrives.[10] Each theory "works" only with a *ceteris paribus* clause, i.e., so long as other conditions remain constant. A scientific theory that works well within our solar system may not function inside a Black Hole or in another stellar system.

What has been even more destructive of our previous assumptions about objectivity, proof and so on is the discovery that, at least at the sub-atomic level, the observer is part of the observed reality; the structure of the reality observed is "changed" by the introduction of the measuring equipment.

We should come back to some of the consequences of these new insights. But before we do that let us take a quick look at the understanding of science in the German-speaking world and then in the Marxist world.

2. *The German language debate:* At least since the nineteenth century, German thought has taken history more seriously than have English-speaking thinkers. Wilhelm Dilthey proposed that the historical method rather than the methodology of the physical sciences should provide the basic framework for understanding reality, since all things exist in history and have their own history.

This is a fundamental issue in Western thought—the reconciliation between the methods of *Naturwissenschaften* and *Geisteswissenschaften* (natural sciences and human or "spiritual" sciences). While the English Radicals (Bentham, Ricardo, Malthus, Mill, etc.)[11] sought to make social science a rational science by reducing all social phenomena to laws (based on laws of human nature, both physical and psychological), the German effort has always been to unite everything by the historical method. Hegel's objective idealism and Kant's subjective idealism uneasily coexisted in the early twentieth century. Kant had effected a divorce between science and metaphysics which Hegel had sought to keep together.[12]

The overcoming of this disjunction between science and metaphysics still remains the central problem of Western thought and the root of its value-crisis. Kant tore apart theory and practice, logic and ethic, the empirical and the transcendental, in the interest of establishing distinct realms for mind, will and taste.

What Kant put apart, Dilthey[13] tried to put together again, through the historical method. Or to put it another way, Hegel had sought to enclose reality in the single concept of Absolute Idea or objective idealism. Three reactions ensued and persist to this day:

(a)   the subjectivist reaction in Kierkegaard and the Existentialists in general shared also by Freud and the Freudians;

(b)   the anticontemplative reaction in Marx, who wanted to keep action/contemplation or theory/practice in a unity and wanted to base it on matter rather than on idea; and

(c)   the Diltheyan reaction which brought the objective world of all cultures and religions within the unifying reality of the individual mind in the act of understanding (*Verstehen*) of experience (Erlebnis).

Historical understanding, as it takes place within the individual mind, became thus for Dilthey the unifying framework for all knowledge; for here the individual mind was participating in the Universal Mind (of Hegel) and gaining access to the dynamic spirit-world in its objective existence containing all realities.

While the Anglo-American world still hoped to make "science" the all-embracing concept of knowledge, the Germanic world preferred to keep science itself as part of the historical understanding, the latter providing the all-embracing framework.

It is important to recognize the consequences of these two tempos in Western thought. The Anglo-American tempo can be characterized as pragmatic, utilitarian, materialist-mechanical, working happily in the laboratory, interested in the physical sciences, untouched by economic, political and philosophical problems; it feels more secure dealing with matter and objects, not people. The other tempo, more characteristic of Germanic thought, is interested in remaking the world by making economics and politics central, and can be termed idealistic. The former concentrates on understanding the given—in order, of course, to use or change it; the latter puts its emphasis on the possible and the ideal, of course in order to change the present towards the ideal future.

The subjective element is more accepted in Germanic thought; while the Anglo-American way has a basic distrust of the subjective. Truth has to be objective.

For the Germans, at least, Heidegger played a large part in enhancing the respectability of the subjective in knowledge. He questioned our prejudice against the subjective. After all, the human *da-sein*, historically conditioned and finite, has to come to terms with the surrounding reality, and the subjective is the essential means for that. Human existence is by nature subjective; there is no need to eliminate the subjective, in the interests of a false ideal of objectivity; without the subjective element knowledge cannot become existentially real. Friedrich Schleiermacher had earlier tried to put the religious element entirely within the subjective, in order to protect it from its cultured despisers who valued the objectivity of rational science. Heidegger made the subjective the locus of all knowledge and understanding, of all will and mind.

Science itself becomes thus respectably subjective; but subjectivity cannot be arbitrary: it has to be subject to critical criteria. This is where Hans-Georg Gadamer comes in with his notion of *pro-theoria*, the ancient patristic-hellenistic notion of foreknowledge as the basis of all knowledge. The art of understanding, of which science is part, is an act of projection of various possible understandings in order that one of these may find confirmation. Understanding is an anticipatory act, projective, seeking confirmation. All scientific theories are thus projections from previous knowledge on to a reality, in anticipation of confirmation by that reality. Such projection of a hypothesis is a tentative prejudgment of what the object could be, a *prejugé* or in simple English, a prejudice. Gadamer puts it this way in his monumental work on *Truth and Method*:

> This recognition that all understanding inevitably involves some prejudice gives the hermeneutical problem its real thrust . . . And there is one prejudice of the Enlightenment that is essential to it; the fundamental prejudice of the Enlightenment is the prejudice against prejudice itself, which deprives the tradition of its power.[14]

Western "rational" civilization likes to be unprejudiced, unbiased, without presuppositions, but alas, that is not possible, says Gadamer. Without prejudice, *pro-theoria*, there is neither science nor understanding. Not only is prejudice unavoidable, but our very prejudice against prejudice comes from a particularly prejudiced tradition—the Enlightenment. The best that reason can do today is not to eliminate prejudice, but to seek critically to discriminate between better and worse prejudices.

A prejudice is not an unfounded judgment; it is a tentative judgment based on previous experience which stands in need of confirmation in the light of further experience. Prejudices in this sense are the major instruments of the scientific enterprise. What science yields are better-tested prejudices.

But how do we critically examine our prejudices in order to see if they are better or worse? Only by making the prejudices with which we operate, themselves, the object of critical understanding. This can, however, be done only by projecting certain prejudgments about what could be wrong with our normal prejudices.[15] In other words, examination of prejudices is itself possible only through projecting other prejudices on to our prejudices. Critical rationality thus always involves the use of prejudgment.

But Gadamer cannot shake himself entirely free of Kant; he suggests that if the process of understanding is based on prejudgments shaped by the understanding person, then in order to understand his prejudices we must look at the historical horizon of that person. It is the structure of his experience which shapes the nature of his prejudices. He has an "effective history", a *wirkungsgeschichte*, which shapes the horizon of his experience and prejudices. "The horizon is the range of vision that includes everything that can be seen from a particular vantage point." People's horizons can be narrow or wide and that affects the nature of their prejudices.

We do not have the space here to engage in a deeper analysis of Gadamer's methodological comments; we should proceed to one of Gadamer's critics,[16] Jürgen Habermas, whose seminal work on *Knowledge and Human Interests* develops Gadamer's work further to propose that there is not only *prejudice*, but also *interest* playing a great role in all knowledge.

Habermas reminds us that a central element in science is methodological doubt. Every conjecture in science must face a refutation. Nothing should be eagerly taken for granted without subjecting it to the cold stare of skepticism. Radical doubt has thus become a basic epistemological category in modern critical theory.

The human consciousness which projects its prejudices and interests to external reality cannot be understood simply through the Kantian categories because when you apply methodological doubt to this category system, you have to ask the question: by what knowing process did Kant know the categories of knowing?

Habermas suggests that consciousness does not exist in isolation nor does it arise ready-made; it is itself the consequence of a social process. That process includes at least three elements, according to Hegel.

(a)   the universal history of mankind;

(b)   the socialization process of the individual;

(c)   the three forms of the Absolute, *à la Hegel*, i.e., religion, art and science, through which human history moves.

It is this concept of consciousness that Marx radically criticized for being too spiritual and contemplative and for ignoring the fact that it is in the process of socially organized labor that the human consciousness takes shape. The process of labor and material exchange, as it takes place in society's interaction with surrounding nature, is itself a consciousness-constitutive process and therefore has epistemological value. It is by labor socially organized that man negates what is given and creates something new, and in that process shapes himself and his consciousness.

Habermas criticizes Marx for reducing the meaning and value of reflection by assimilating it to a subsidiary role in socially organized labor. Habermas would insist that productive activity is not sufficient for self-generation; critical revolutionary activity based on correct theoretical reflection is also necessary for constituting oneself.

Here we come to one of the crucial problems of our

scientific method. Schelling had already in 1802 fought against the reduction of all knowledge to practical knowledge. This abhorrence of theory and speculation is dangerous. *Theoria*, for the Greeks, meant movement from the manifold and the particular of things, to the unifying ideal of the Logos.

That kind of theoria may be difficult for us today. All we can hope to have today is critical theory, i.e., the theory that constantly uses methodological doubt to examine all our accepted prejudices in order to see whether better ones cannot be conceived, to see what class interest shaped one particular prejudice at one particular time, and so on.

We should come back to some of these questions; before we do that, however, we should take a quick look at the Marxist perspective on science.

3. *Marxist Views of Science and Understanding*: The Marxist intellectual would criticize both the Anglo-American and the German discussions as being overly abstract and reflection-oriented. Gadamer can be definitely accused of assuming a framework of thinking which takes the consciousness of the individual and its relation to a world of objects as central. Habermas, as an ex-Marxist of the Frankfurt School saw this problem and sought to correct the individualist emphasis of Gadamer with an analysis of the class origins of each consciousness.

The debate between Frankfurt School thinkers like Adorno, Horkheimer and Habermas on the one hand and Orthodox Marxists on the other have to do with the importance given to (a) reflection over against socially organized labor, and (b) the function of the critical philosophy as methodological doubt of all received positions.

This has to do with one's conception of Truth. For the Marxist, "true knowledge must reveal the logic of the evolution of social being",[17] not just one individual mind's perception of external reality. But for the modern Soviet academician, Truth is *both* a process in which the infinite material world with all its regularities is reflected in consciousness, *and* man's reshaping of that world through sociohistorical practice.

In Western liberal thought one sees both the tendency to make the individual primary and society secondary, and also to make reflection primary and practice secondary. Western liberal thought would thus be criticized by the Marxist as too individualistic and too psychological.

Marxist theory, however, holds on to the "copy" or "reflection" theory of the relation between the mental world and the external world. This view of truth as *"adequatio rei et intellectus"*, which is common to Gadamer and the Marxists, as well as the Thomists, is one that needs fundamental questioning for it seems to be the key to a radical reinterpretation of reality.

The main difference between the Marxists and Gadamer would, however, be in the realm of the definition of the *practice* which is necessary to confirm a conjecture. For the pragmatic Anglo-Saxon it is basically the "experiment", the empirical ritual in the laboratory. For the Marxist

> . . . practice is the socio-historical activity of people: activity in the sphere of material production, in the sphere of the class struggle and social relations, in the sphere of scientific observation and scientific experiments, which depend on the corresponding level of material technology.[18]

For the Marxist what confirms knowledge is not mere laboratory experiment, but rather the social experiment, which includes laboratory experiments, but is more concerned with the struggle for constructing a new political economy.

While Western scientists are now beginning to let go of their former claims to objectivity, the Marxist scientist still insists on "the objective character of scientific knowledge, its reflection of an objective reality existing independently of the subject";[19] though this does not involve any denial of the role of subjectivity in knowledge.

One thing which is impressive in the Marxist philosophy of science is its effort to keep science and philosophy integrated.[20] The physical as well as the human sciences are seen in an openly philosophical context, and the conclusions of sciences constantly change philosophy and keep it growing. Western readers have been conditioned to see only the negative aspects of this ideological domination of science by ideology; they often cite Lysenko's genetics or Stalin's linguistics as examples of the consequence of this domination.

In the West one makes an ideal of the independence of science from ideology. The net result, however, seems to be that science, by pretending to ignore the political economy within which it functions manages merely to reinforce and strengthen the structural anomalies and injustices in society. The Soviet academician consciously accepts the relationship between science and philosophy.

> Under certain circumstances the natural sciences have an ideological function. There is no such thing as bourgeois (or communist) physics, chemistry, etc., but there are various ideological interpretations of the major discoveries made in the natural sciences.[21]

Dean Arthur Peacocke of Claire College (Cambridge) has insisted that there is no African physics or Asian chemistry, but drew the dubious conclusion therefrom that cultural conditions were not decisive for the shape of scientific development. Peacocke's view was probably a fragment of the old positivistic view that in science there can be only one true view of any given set of facts. More common in the West is a sort of easy pluralism which says that between various possible interpretations of a given major discovery there is no need to choose. Not to chose is to choose the given.

Oizerman insists that to philosophize is to choose, and not to philosophize, even in science, is to philosophize wrongly, proceeding on the basis of unexamined assumptions. Not to philosophize about the meaning of science is to fall victim to our oppressors and exploiters who do not want us to recognize the source of our oppression and exploitation.

Marx would say that the West still follows the philosophers who would explain the world but do not desire to change it. Science and technology, for the Marxist, exist in order to serve in the process of socially organized labor for the welfare of the people. Reality is not *theoria*, but *praxis*; humanity using *theoria* to deal objectively with surrounding reality in order to transform reality and humanity itself in the process. We constitute the world and ourselves, not by thinking, but by interacting with the material world of which we are part. Reflection is not truth, but something necessary in the interaction between man and surrounding reality.

These processes of human interaction within society and with nature have their own intrinsic laws—these are the laws with which the science of political economy deals. These laws, however, are not static or given. Marx acknowledges his indebtedness to Hegel for both of his key ideas—labor as constitutive of man, and knowledge as always dynamic and changing. As he stated in his *Critique of the Hegelian Dialectic and Philosophy as a Whole* (1844):

> The outstanding achievement of Hegel's *Phaenomenologie* and of its final outcome, the dialectic of negativity as the moving and generating principle, is thus first that Hegel conceives the self-creation of man as a process . . . Hegel's standpoint is that of modern political economy. He grasps *labor* as the *essence* of man, as man's essence which stands the test.[22]

But Marx would criticize Hegel for recognizing only abstract mental labor as labor constitutive of man. "Labor is man's *coming-to-be for himself* with *alienation*, or as *alienated* man."[23] Hegel made the mistake of taking alienated human labor, or the abstract thinking of alienated human beings as genuinely constitutive of humanity.

For Hegel man equals self-consciousness. It is in his consciousness that he has to overcome alienation. It is through labor that man re-enters into proper relationship with external objects, by seeing in one's consciousness their true nature. The labor that constitutes the human person in his authenticity, however, is intellectual labor, and philosophy is the means of constituting oneself, for Hegel.

It is against this view that Marx has reacted. Hegel's

view makes it possible only for the intellectual elite to become authentically human. Marx's contribution was to see theoretical reflection as integrally related to the praxis of socially organized labor, and meaningless without that integral relation. Hegel saw the universal mind reflected in the individual mind. Marx would prefer to see it reflected in the corporate social mind. The intellectual, by himself, does not have access to truth; nor does science divorced from political economy. It is necessary to integrate the intellectual and the working class, as well as individual and society, but both these integrations have to take place in an ideological paradigm that combines science and philosophy, technology and political economy in a single framework.

Marxism alone, among contemporary philosophies, has such a framework or paradigm which can integrate the physical and the human sciences, science and philosophy, technology and ideology into a single integrated paradigm.

Neither the Christian faith, nor any of the other religious or secular non-Marxist ideologies, has, so far as I know, succeeded in providing such a framework. To these religious frameworks and their inadequacies we must now turn before we try to delineate the contour of a new paradigm.

CHAPTER TEN

Towards a New Paradigm for Reality

A paradigm is a framework, a *gestalt*, a pattern that we project upon reality in order to perceive it. Paradigms are always formed from previous experience and theoretical reflection. Paradigms may be consciously held or unconsciously assumed. Without a paradigm or pattern projected by the mind, perception seems impossible.

From a Christian perspective, it may be stated that paradigm change is the key to spiritual growth. A spiritually mature person or community is one that is able to project a pattern that yields a good sense of reality, about what is good or evil for oneself as well as for others.

CRITIQUE OF EXISTING PARADIGMS

Our commonly held paradigms may be described as theistic, atheistic, agnostic, deistic, secular-dialectical, secular-liberal and so on. This aspect, however, relates only to one dimension of the paradigm which has to do with the question of a description of how the whole cosmos hangs together.

The theistic perspective usually holds that God, who is distinct from the cosmos, is the Creator and Sustainer of the universe. The atheistic paradigm boldly denies that there is such a God who created and now sustains the universe. The atheist would have to assume that the universe or matter is self-existent, self-caused and self-sustaining without any outside causation or control. These are exactly the assumptions that the theist makes about God as Creator.

The agnostic perspective modestly holds the view that the issue between theism and atheism cannot be resolved, and also that it is not necessary to resolve it. One can concentrate on how to live in the world rather than on origins and causes. This position generally leads to a pragmatic paradigm which holds that reality is that which works sometimes to my advantage, sometimes not so.

The secular-liberal position is not so much agnostic as practically a cross between atheism and deism. It is not necessarily atheistic in the sense of a passioned opposition to belief in God. In fact it is permissive or pluralistic, i.e., it holds the view that people may or may not believe in God but that in actual practice it makes no difference. Pluralism, often so lauded by liberalism, becomes a form of indifferentism—the view that differences do not matter. Some forms of the secular-liberal position claim to be religious. They would, for example, concede that God created the world, but then state that God has now left it to human beings to shape it, by learning its laws and thereby controlling and directing its development. In effect this amounts to more or less the same as the Deist perspective which holds that God created the world, gave it certain "natural laws" and left it to itself.

RELIGIOUS PERSPECTIVES—WEST ASIAN AND EAST ASIAN

It is an often unnoticed fact that all world religions come from Asia. West Asian religions are, generally speaking, based on a personal God, transcendent, Creator, distinct from the universe and man; East Asian religions generally conceive the universe as an emanation from God, an extension (body) of God, or identical with God, or in some cases, brackets the whole problem of God as irrelevant (Buddhism).

But none of these religions has given us an adequate and comprehensive paradigm within which to understand all reality. There is some basis to the claim that in general East Asian religion can be more easily compatible with science, and that there are some similarities between the cosmology of East Asian religions and that of modern physics. This has been rather impressively argued in Fritjof Capra's *The Tao of Physics*.

It seems clear that the West Asian religious perspective needs to be balanced by a deeper knowledge of East Asian philosophies and spiritualities. Judaism, Christianity and Islam all have glorious philosophical heritages, but these are largely neglected today, under the impacts of secularism and pragmatism. They should themselves revive their own philosophical tradition, and then come into a three-cornered "dialogue" among (a) West Asian religious philosophy, (b) East Asian religious heritage and (c) modern science/technology and philosophy of science.

In general, Buddhist philosophies, precisely because they do not presuppose any belief in God, appear to be more compatible with a secular cosmology. Among the many schools of Buddhist philosophy, Indian or Chinese, Japanese, Tibetan or Sri Lankan, the *Mādhyamika*[1] school is probably the most interesting. *Mādhyamika* philosophers generally agree in refuting the assumption that something really exists—either things or ideas or God. The *Svātantrikas* (a division of the *Mādhyamika* school) were prepared to concede that things existed as being-things, or as self-evident sense-impression creators, but this was not sufficient ground to assume their real existence. The other *Mādhyamika* school, the *Prāsangikas*, embarked on a kind of Wittgensteinian language-game. They abandoned the temptation to explain reality. They pointed out, like the Sceptics among the Greeks, the inherent shakiness of every logical or verbal postulate. They would not accept any kind of reductionism, and developed a formidable repertory and technique of logical analysis by which they could refute any given postulate or proposition. But *Prāsangika* philosophy is more than a mere language game. Their thinkers were not concerned with developing a new speculative philosophy but were seeking some meaning for existence. This they did, however, by developing a highly sophisticated epistemology.

Denying the notion of an "essence" of things, rejecting the view that things exist by virtue of a constitutive principle through which they are what they are (the *Svātantrika* view), the *Prāsangika-mādhyamikas* claimed that a judgment of perception about what is under consideration comes about in a person by epistemic conditions alone.

The *Prāsangikas* delight in demolishing other people's theories about essence and existence, function and causality, being and nonbeing, substance and quality. They admit only relational existence and everything comes into being only in this relational way and not as discrete entities existing in themselves independent of such relation. This notion of *Pratītya-samudpāda* or *Conditioned Coemergence* belongs to the heart of Buddhist philosophy and is worthy of further study by anyone interested in a modern cosmological paradigm. The originator of the *Mādhyamika* school, Nāgajuna (second century A.D.) himself wrote a work called the *Heart of Pratītya-samutpāda*[2] and the very first invocatory *ślokas* of his *Mādhyamika-Kārika* describe the concept of *Pratītya-samudpāda* in terms of four pairs of negatives—neither coming to be nor ceasing to be; neither permanence nor impermanence; neither unity nor diversity; neither coming-in nor going-out.[3] Nāgājuna does not say either that things exist or that they don't. Things have only an interrelated existence —not each thing in itself. At the pragmatic level or *samvrti* one can act as if things were real, but at the transcendental level or *paramārtha* there is only *śūnyata* or the Absolute as nonbeing.

Science at present deals only with *Samvrti-satya* or pragmatic truth, but already there are indications within it that the *samvrti* level is but an initial level of apprehension of truth, always pointing to a transcendent level beyond. We have to pass through the *samvrti* level to get to the *paramārtha* level.

The *Pratitya-samutpāda* view that the cosmos has only a relational existence through conditioned coemergence is itself something that goes beyond the *samvrti* level, but it refers to the reality apprehended at that level. Seen from the *paramārtha* level, this conditionally coemerging universe is only *śūnyata* or the Absolute as nonbeing. The apprehension of this *śūnyata* or nonbeing is the ultimate experience or *nirvāna*.

The *Prāsangika* school branched out of the *Mādhyamika* school of Nāgārjuna (second century) and Aryadeva (third century); Buddha-pālita at the beginning of the fifth century A.D. gave birth to the Prāsangika school by developing sharp *reductio ad absurdum* arguments against all commonly accepted conceptual formulations of cosmology and epistemology.

The *Mādhyamika* school spread to China early and shaped its civilization for some eight centuries. Kumārajivā (fl. 405 A.D.) was himself half Central Asian (from Kuci beyond the Pamirs in Chinese Turkestan), studied in Kashmir and went back to Kuci. From there he was taken prisoner during a Chinese invasion of Kuci and carried to China as a trophy of war! There Kumārajivā, with the patronage of the King, undertook a monumental translation project for producing Chinese versions of some 300 classical Indian Buddhist texts, mostly from the *Mādhyamika* School. He was equally expert in Sanskrit and Chinese. It was thanks to these Chinese translations that we still have access to much of Nāgarjuna's thought.

Kumārajiva (in Chinese Ciu-mo-lo-shi), Paramārtha (Po-lo-mo-tho), Dharmabōdhi (Ta-ma-phu-thi), and other Indian teachers spread the teaching of Mahayana Buddhism in China and Central Asia including Tibet in the sixth century A.D. This philosophy needs today to be revived as a philosophical medium for enriching any comprehensive paradigm that we may wish to devise for advancing and integrating present scientific knowledge.

Several Western scholars and scientists who set out

on this path—Oppenheimer and Einstein are only two examples—have not been able to advance very far, and it seems like the creation of a new paradigm would require sustained interaction between scientists and philosophers from East and West for quite some time. Some wealthy foundation or individual should set up a project to put together learned Chinese, Indian and Western scholars of philosophy and science to seek the contours of this new paradigm that can show us the way to the future.

BUDDHISM, HINDUISM AND TAOISM

Buddhism, though of Indian origin, has today become universal, permeating the cultures of China, Japan, Korea, Kampuchea, Laos, Vietnam, Sri Lanka, Burma, Thailand, Tibet, Ladakh, Mongolia, Nepal and Indonesia, as well as the Central Asian republics of the Soviet Union. It has recently spread to the West also. Hinduism, on the other hand, had remained largely an Indian religion with some pockets in Indonesia; in modern times it has gained many millions of adherents in the West and is becoming universal. The Tao, Chinese in origin, is also becoming universal in our time, and has been fast winning converts in the West.

All these three religions have philosophies just as profound and as illuminating as the classical or modern philosophies of the West; but the former do not carry much appeal to the literate people of our time. One can easily be regarded as a learned philosopher or theologian in either West or East without serious acquaintance with the texts of these religions. Christianity has been largely responsible for this closing of doors to the full wealth of the heritage of humanity and for developing a civilization that is as parochial and arrogant as it is insular and uninformed.

There are two aspects to the religious philosophies of the East—the astute dialectic of their philosophical logic and the deep and satisfying wealth of their religious experience and perception. These two aspects can be easily separated, and that is the weakness of Oriental philosophy as now taught in universities in the West and in the East.

If science is to advance the welfare of humanity, there must be a genuine encounter between those philosophies and spiritualities of the East in their full spiritual-intellectual vigor, and the sophisticated though yet spiritually arid philosophies and science/technology of the West.

This encounter is unlikely to take place in a single mind. There are several people in India—several does not mean many— who have tried to achieve this synthesis. I can readily think of three people who are both philosophers and scientists, coming from three different religious traditions in India. First, I think of Dr. D. S. Kothari, formerly Chairman of the University Grants Commission in India, who writes with a deep knowledge of modern science, from the perspective of the Jain religious tradition of Mahāvira. His passionate interest, not only in science and the humanities, but also in spirituality and personal character, has led him to undertake serious efforts to bring religion and science into a single perspective. His advanced age and failing health may prevent something really dramatic coming out of these heroic efforts.

A second person in the same field is Dr. Sampooran Singh, D.Sc., head of the Central Defense Laboratories in Rajasthan, who has already published several works in this field. His background is that of the Sikh religion of Guru Nanak and the later Gurus and the Granth Sahib. Once again, one sees how great the task is and how unable a single mind is to cope with the vast range of problems posed by science and the philosophy of science on the one hand and Eastern religious philosophies on the other.

A third person, much younger, a well-known scientist in his own right, is Professor E. C. G. Sudarshan, whose background seems indeed exotic. Born into an Oriental Orthodox Christian family (my own Church) he has become a convert to Hinduism and has acquired deep perception in Hindu spirituality and philosophy. Professor Sudarshan is attached both to the University of Texas and to the Indian Institute of Science in Bangalore, and is known for his contribution to modern physics in theorizing on *tachyons* or subatomic particles traveling faster than light, as well as on creating new equations which adapted the rigorous theory of partial coherence in classical optics (Emil Wolf) into a quantum framework.

In a country like India there should be many others who are in pursuit of the unifying paradigm that integrates the scientific perspective on reality with the religious one. There must be some also among the millions in China, Japan and other Asian countries.

What we need is a mechanism that will bring these different minds together for a concerted and sustained effort in search of a paradigm. I would like to insist that the East Asian religious perspective is one that we cannot afford to ignore. This perspective has been grossly misrepresented in the West by even such well-known writers as Albert Schweitzer. For example, when the Buddhist philosopher advances his *Śūnyatā-vāda* about reality, many Western scholars misunderstand it as a contention that reality does not exist, as a kind of nihilism. One fails often to see the difference between "no-reality" doctrine and "no-doctrine about reality" doctrine, as Professor T. R. V. Murti has shown.[4] Neither do many realize that when Śankara speaks of Māya, he is speaking of the conditioned nature of reality perception and not about the world being an illusion. The world is real, but not as it appears to us. Reality is veiled and knowledge has to unveil it, discover it as it is. When in the ultimate experience of knowing, reality thus unveils itself, one sees that Brahma, self and world are not three disjunct realities, but in fact one. Philosophy or perception of reality is then based on the experience of this supreme knowledge, not learned from books or propositions. True knowledge is beyond the subject-object kind of empirical knowledge where Brahman has nothing to do with the subject knowing the object and where all three remain distinct and distinguishable.

Both Hindu *Vedānta* and *Mādhyamika* hold to the dialectical apprehension that all conceptually-grasped truth in which there is consciousness as subject and consciousness of something as object, is originating in a conditioned mode, and is not the ultimate truth. This view has affinity with that of Kant and the neo-Kantians in the West, but is not the same. The concept and the world arise simultaneously under certain conditions which can be transcended.

Science is already at the door of this perception. We know now that while we can introduce independent standards for measuring and checking reality, it is not possible to eliminate the subjective in scientific perception. We have begun to see that all knowledge is relational and not absolute. We are beginning to see that the act of knowing is a constitutive act and shapes the knowledge yielded. We know also that the relationship of ego and consciousness is highly problematic as Kant and Sartre as well as Heidegger have adequately shown. We are beginning to see also that all reality is one interrelated system.

If one takes seriously what has been said about faith as nurture and support rather than as encounter, then we can begin to see that transcending the subject-object dichotomy is necessary also for true faith. A proper philosophical-theological perspective should reveal to Christians also that they have been apprehending their faith intellectually on highly questionable philosophical grounds which separate God, Man and World as three disjunct realities. Christians can come to a deeper and philosophically more adequate grasp of their own faith through proper and profound study of Eastern religious philosophy in Hinduism, Buddhism and Taoism.

And since Christianity is still a pervasive influence in inhibiting Western thought, the liberation of Christian theology from the shackles of doubtful metaphysical assumptions which obscure the very nature of Christianity and from the endemic reluctance in Christianity since the Middle Ages to philosophize deeply, could lead to a liberation for Western civilization as such. Here both science and the Eastern religions as well as the perspective of Oriental semitic Christianity have a large role to play.

The inhibitions of the West in this regard are best exemplified in the failure to extend recognition to C.G. Jung, the Swiss psychologist, who is one of the few in the West to begin to penetrate the heritage of East Asian religions. He saw clearly the problem of *causality*, the central interpretative principle in modern science. His search for acausal interpretations of reality began with the study of East Asian religions. As he stated in the first chapter of his work on *Synchronicity*, "the connection of events may in certain circumstances be other than causal, and requires another principle of explanation."[5] Usually we resort to the lazy method of ascribing to chance what we cannot explain in terms of causality. Jung, exploring further a view opened up by Schopenhauer (who was also an ardent though not quite successful student of East Asian religion), was dissatisfied with chance as an explanation for the causally inexplicable.

Jung turned to China and Taoism and found principles in the *I Ching* "for grasping a situation as a whole and thus placing the details against a cosmic background —the interplay of *Yin* and *Yang*".[6] It is precisely Jung's interest in astrology and *I Ching* that scared many pious Westerners and turned them off from even listening to Jung. But he found in Taoism a psychological approach to reality that was just as illuminating as the Western principle of breaking up everything into cause and effect in a series and not wholistically.

For Lao-Tzu, the Taoist philosopher, Tao is also the Absolute as nonbeing. Nonbeing does not mean the absence of something, but rather like the empty space inside a vessel or within the frame of a door, the emptiness is the most meaningful and useful. As the Tao Teh Ching puts it:

Unmeasurable, impalpable
Yet latent in it are all forms;
Impalpable, unmeasurable
Yet within it are all entities;
Unclear it is and dim.
Ch: XXI

The central point of Tao is again the overcoming of the subject-object dichotomy and seeing their unity within the whole. The problem as well as the point of modern science is the attention it pays to the empirical and to the detail. Our way of thinking helps us to see the shape of the vessel or the elegance of the door-frame, but not the meaning of the space contained in each. Modern science has taught the West to forget what Philo and Hippocrates and Pico della Mirandola taught them about the unity of all; but it is precisely that science, yes, modern physics, which now tells us that all things are interconnected, and nothing really exists as discrete and separate. But the integrating paradigm has yet to emerge.

A CRITIQUE OF SECULAR PARADIGMS

We have mainly two general sets of paradigms in Western culture—the pluralistic, liberal, secular, Western scientific paradigms with their market economy system on the one hand, and the more closely integrated Marxist paradigm with its supposedly socialist system on the other.

The liberal paradigms are in a framework which is basically antidogmatic and antiphilosophical in tempo; practice and use are its more immediate concerns—individual use, family use, use of corporations and states. This framework is satisfied with the thought that an integrating framework is not only not possible but also not quite necessary. There are some who still strive for a general theory of relativity (GTR), but to most scientists, a little mechanistic paradigm, a little quantum-mechanical paradigm, and a little special theory of relativity paradigm together meet the main needs.

As for an ideology, the West would pretend to be content with a critical liberalism—an aversion to all dogmatism and tradition; a need to question every assumption and every opinion except those whose questioning would upset one's interests, a broad tolerance that lets a hundred flowers blossom; a general view that "you may be right, he may be right, and I may be right; it does not matter too much so long as he or you do not threaten my security, my comfort and my conflictless respectability."

Liberalism has little positive content. It is quite happy

to acknowledge a few general principles like the personal freedom of the individual, the dignity of the person, the need for justice, and a broad tolerance of dissent. Liberalism reveals its weakness when confronted either with a more resolute and self-confident ideology or with catastrophe on a large scale. In fact it is so confronted at this time—both by the self-confident power of Marxism and by the fundamental sense of catastrophe today endemic inside industrial-capitalist civilization. So liberalism itself becomes intolerant. It is well-known how difficult it is today for a Marxist party or Marxist individuals to survive and function within the so-called "free" societies of the developed market economy system. Even in those countries like France and Italy, with strong Communist parties, there is a thick wall of separation dividing the Marxists and the liberals, so that it is difficult to find full social acceptance for a committed Communist in non-Marxist circles. It is the industrially advanced liberal society that is most prone to the doomsday psychology of fearing impending catastrophe either in the form of a nuclear holocaust or the outbreak of a Third World War or even one caused by the melt-down of several peace-time nuclear power plants. It is in these countries that one sees more of the fear of ecological catastrophe—resulting from air and water pollution, from climatic change due to the greenhouse effect produced by excessive carbon-dioxide in the atmosphere, or from total disruption of the eco-system through upsetting of the eco-balance that maintains life on our planet.

Liberalism as an ideology seems unable to cope with the big issues; neither does it seem capable of providing integrative paradigms. Critical rationality remains a highly useful tool in the hands of humanity; but it shows signs of not being able to stand on its own feet. It can be more fruitful within a more integrative paradigm, but it cannot itself produce that paradigm. The bigger problem with liberalism is that it seems to be so integral to the World Market Economy System, to a critique of which system we must now turn.

THE LIBERAL IDEOLOGY'S MAJOR PRODUCT—
THE MARKET ECONOMY SYSTEM—A CRITICAL COMMENT

The market economy system has been able to achieve a considerable enlargement of the middle class, multiply the numbers of millionnaires by several thousands, and bring a tolerable (from the perspective of the Two-third World) level of income to the working class. It has also brought some dignity to labor in the industrially developed countries; and the poor who had been formerly regarded as the scum of the earth can today have good clothes, cars, houses and so on.

But the problem of gross inequalities of income and frighteningly high rates of unemployment remain the endemic problem of all market economy societies, whether industrially advanced or backward. And these societies do not seem to have any real plan to overcome these two problems, both of which are of primary interest to the poor. The poor and the oppressed of the world should not then be blamed for being foolish or unwise if many of them have a greater sympathy for those economies where these two problems have been resolutely handled.

But there are other reasons why not only the poor, but all friends of justice prefer the socialist economic system to the so-called liberal-democratic or market-economy ideological pluralism. The reasons may be stated as follows:

(1)   The industrialized market economy countries, including ostensibly noncolonialist nations like Sweden and Switzerland have inherited and benefit from the unequal and exploitative economic system built up by American, European and Japanese colonialism and imperialism—built up through the last several hundreds of years. They still regard this unjust and cruel system as a framework for international economic relations in which they can continue to have major benefits at the expense of the poor of the world. To be committed to such a system is to be committed to the perpetuation of injustice.

(2)   Some of the more powerful industrialized market economy countries often openly seek to prevent the less developed countries from solving their own problems by refusing to help them with the technological know-how and the capital assistance needed to start certain key basic industries like steel or nuclear power. India, for example, was cold-shouldered by the U.S.A., the U.K. and West Germany when she first sought assistance, soon after national independence, to build steel mills. It was only after India began constructing steel mills with the assistance of the Soviet Union that these nations also began to offer assistance for the construction of steel mills. The chaos and catastrophe caused at the Tarapur nuclear plant by the refusal of the U.S. Government to fulfill its contract obligations in the matter of supplying enriched uranium also illustrate the tendency of the industrially advanced market economy countries to keep the less developed countries technically and industrially backward and heavily dependent. One could give scores of other examples like the intrigues against developing oil refineries in the Middle East, the dirty tricks in oil prospecting in Asian and African countries, and so on. To be committed to such a system is to be committed to dishonesty, oppression and exploitation.

(3)   It is also true that the general policy of the industrially advanced market economy countries is to keep the Two-third World firmly anchored to the world market economy system, and they use every trick in the bag and all forms of force to prevent any of the less developed countries from breaking out of the system. By ideological brainwashing of the intelligentsia, through newspapers, news agencies and all media-related activities as well as through dirty destabilization tricks, and even through "best-seller" literature, the powerful neocolonialist market economy system seeks to keep the minds of the Two-third World enchained to itself, and to keep control not only of the economic process but also of all intellectual develop-

ment. To be committed to such a system is to be committed to the prolongation of our spiritual and intellectual enslavement.

(4)   One of the means of keeping the Two-third World enslaved is to attempt maximum possible stimulation of market economy private enterprises in the Two-third World through financial agencies like the World Bank and the International Monetary Fund and through the World Banking System. These agencies maintain, propagate and often enforce a negative attitude towards the state sector, or where this is not possible, coopt the state sector in order to make it serve the private sector, as is largely the case in India today, where subcontracts given to private enterprises neutralize all the good that the state sector could do for the common people. The state sector now becomes another fat cow to be milked by the entrepreneurs or at times even the catalyst that stimulates private enterprise. In this manner this market economy system and ideology coopt what is known as the "public sector" into a major instrument for serving private interest. Its financial power is almost always used to reinforce itself, and to be committed to this system is to be committed to the vested interests of the privileged few.

(5)   The world market economy system effects a steady decline in the Two-third World's share in international trade, squeezing out the sixty percent of world population from any major share in the so-called benefits of the market economy system.[7] This they achieve through a more "rational" use of science and technology, through greater integration of the economic life of the leading market economy countries, and through artificial increase of consumption in the already affluent countries, along with stringent restrictions against the import of goods into these new markets from the less developed countries, and a policy of agricultural and raw material protectionism which aims at reducing affluent country dependence on the

less developed countries. To be committed to such a
system is to be committed to the greater impoverish-
ment of the poor.

The world market economy system which the liberal ideology has created becomes a major instrument
of oppression and exploitation also through unfair
trade terms. While the costs of spiralling inflation in
the advanced industrial countries of the market economy world are charged to the less developed countries
in terms of the highly increased prices they have to pay
for what they import, they receive no major increase
in price for the goods (except oil) they supply to the
affluent. This was the case even before the present two-
digit inflationary spiral began. In absolute terms the
purchasing power of the less developed countries fell
by at least $6400 million between 1955 and 1970.

At the same time the income taken out by foreign
investors directly from the less developed Two-third
World grew enormously.[8] Between 1960 and 1970
that increase was 136 percent. To be committed to such
a system is to be committed to the plunder system
established during the colonial era.

(6)   The market economy system created by the liberal
ideology has now institutionally entrenched itself
through the system of Trans-National Corporations,
by which capital from the advanced industrial countries can, in league with local entrepreneurial interests,
establish themselves in the Two-third World economic
and intellectual development. These corporations appear to be useful in catalyzing industrial development,
in increasing capital and know-how available to the
Two-third World, and in advancing the development
of science and technology in the less advanced countries. But in fact the income taken out of the Two-third
World countries by these corporations has been consis-
tently higher than whatever they have supposedly
brought in. They have put the countries of the Two-
third World into a deep trap of economical and technological dependence, shackled and burdened them

with heavy chains and loads of foreign debt, and have intellectually and spiritually castrated the cultures of these countries. To be committed to such a system is to acquiesce in the strengthening of the shackles that bind us to our exploiter and oppressor.

(7)   Two major instruments by which the Market Economy System, instead of becoming more humane, seeks to strengthen its stranglehold on the Two-third World, are the phenomenal increase in the Arms Trade and the Know-how Trade.

The Two-third World accounted for only four percent of total world military expenditures (excluding China for which precise figures are not available) in 1957, according to the SIPRI Yearbook 1978. In 1977 the figure rose to fourteen percent or, in absolute terms, from $17,425 million in 1957 to $70,300 million in 1977 at 1957 prices. The value of imports (at 1975 prices) of major weapons by Two-third World countries rose from $1202 million in 1957 to $8161 million in 1977. This enormous expanding market is used not only to exploit the Two-third World, but to increase the military as well as economic control of the Two-third World by the developed countries. It also helps to reinforce the entrenched power of privileged and vested interests in the Two-third World.

The sale of technology is an equally frightening development in the Market Economy System. Daniel Bell, in his description of the new "Post-Industrial Society", lists five aspects of this transition from Industrial to Post-Industrial Society.

(a)   In the economic sector, there is change of emphasis from producing *goods* to producing *services*.

(b)   Occupationally, the Post-Industrial society puts its premium on the professional and the technologist, thus stressing again knowledge and skill.

(c)   In research, the axial principle of innovation emphasizes again the centrality of theoretical knowledge.

(d)   In planning, the control of technology gets central place.

(e)   Even in decision-making, it is the new intellectual technology of "human engineering" that gets central place.

Small wonder then that out of the 2,978,204 scientists and engineers today engaged in fundamental research, only 26,891, or less than one percent of the total, are in Africa, and only 39,603 or 1.33 percent are in Latin America. North America spent $35,978,815 for research and development (1974 figures) and Western Europe spent $24,212,659, the two together spending almost sixty percent. If one includes Japan and Oceania, the percentage comes to nearly seventy. The developing countries' total share of total research and development expenditure was 2.6 percent while the Socialist countries spent 28.27 percent of the total.

One sees clearly that in this oppressive system science and technology become a major instrument of exploitation. Science and technology cannot come into their own when sixty-three percent of the world's population has only a 2.6 percent share in the world's research and development funds. If science and technology is to develop for the benefit of man there is no other way except to reorganize the world economic structures into a more equitable and just system.

Science and technology first developed in the Market Economy System. The Socialists came in later, and have made considerable headway. Of the total number of about three million scientists and engineers engaged in fundamental research and development, the U.S.S.R. alone has 1,169,700, or 39.28 percent of the total, and other East European countries have 324,462 or 10.9 percent of the total. Between them they have fifty percent of the world's engineers and scientists.

Some people think that both the Socialist bloc and the market economy bloc are equally exploitative. This is far from the truth. It is true that often the market economy bloc technology is more advanced than that of the Socialist bloc. But Socialist technology can be made available mainly for the advancement of the lowest income sectors, while market economy bloc technology, operating through Trans-National Corporations, works more to the benefit of the entrepreneurial and managerial classes, and becomes more clearly an instrument of exploitation and dependence creation.

Socialist technology also creates some dependence relations, but is definitely less exploitative. Socialism itself is not completely free from imperialist tendencies, and this constitutes one of its major weaknesses, which we in Asia know well. Yet it seems clear that science and technology for the benefit of humanity is more likely to grow within a socialist rather than in a market economy or in a mixed economy like ours in India.

A CRITIQUE OF CURRENT SOCIALIST IDEOLOGIES

While between the Market Economy System and the Socialist system the choice falls unmistakably on the latter, the present reality of the socialist systems is itself not beyond criticism.

The brief critique here cannot deal with all aspects and all types of Marxist philosophy. We can only make oblique references to the Euro-communism of Maoism as variations on the main theme, and have to direct our attention more to its official ideology in the most developed socialist country—the Soviet Union. We refrain also from the Gulag Archipelago type of criticism.

The fundamental question relates to the validity of knowledge or epistemology. How are concepts related to facts? How is Being related to consciousness? The Market Economy West plays with the Correspondence, Coherence and Pragmatic criteria for Truth. Taking it for granted that truth is propositional, they can be too easily satisfied with quite unscientific and arbitrary definitions of truth. Kant and the Phenomenologists at least saw the problem of the subjective-objective dialectic in all knowledge, the dialectic between being-in-consciousness and being-for-consciousness, between the *en-soi* and the *pour-soi*. But most empiricism in the West still tries to skirt the issue.

In Marxism, it was Lenin, very much a philosopher in his own right, who elaborated a simple "copy" theory of the relation of mental percepts to external objects. Today the Marxist position is much more sophisticated. Lenin recognized that "no natural science and no materialism can hold its own in the struggle against the onslaught of bourgeois ideas and the restoration of the bourgeois world outlook unless it stands on solid philosophical ground."[9] But Lenin himself was too heavily dependent on the Hegelian Dialectic turned upside down by Karl Marx. He saw only three alternatives—materialism, idealism, or skepticism—and he wanted the dialectic of Hegel in a materialist framework; this was a clear choice for Lenin—both idealism and skepticism are reactionary; only materialism is progressive and scientific. Philosophy itself is scientific, not anything extraneous to science, according to Engels as well as Lenin.

Academician Fedoseyev, in an article entitled "Scientific Cognition Today: Its Specific Features and Problems," puts it thus:

> Dialectical-materialist philosophy does not deny the role of formal-logical methods of research, social factors and individual creative activity in the process of cognition. But it shows the significance of these factors in relation to what constitutes the real essence of human cognition—the interaction of the subject and object in the process of practical activity. This interaction is interpreted and explained in Marxist philosophy on the basis of acknowledgment of the materiality of nature and society, the dialectics of objective reality and the reflection of the latter in consciousness, on the basis of the principle of the social character of cognition. Both the thought and practical activity of man are determined by the laws of objective reality. Man's subjective activity is not absolute and arbitrary; in the final analysis it is determined by external reality, by objective dialectics.[10]

This is a much more sophisticated position than the copy theory developed by Engels and Lenin. Engels, an outstanding philosopher, sometimes was quite simplistic on epistemology:

> Contrary to idealism, which asserts that only our mind really exists,[11] and that the material world, being Nature, exists only in our mind, in our sensations, ideas and perceptions, the Marxist materialist philosophy holds that matter, being, is an objective reality existing outside and independent of our mind; that matter is primary, since it is the source of sensations, ideas, mind and that mind is secondary, derivative, since it is a reflection of matter, a reflection of being.[12]

The expression "copy theory" gives place in later Marxist literature to the notion that external reality is reflected in the mind of man. As another Academy of Science Volume, (*The Fundamentals of Marxist-Leninist Philosophy*),[13] puts it in less technical language:

> Materialism in the theory of knowledge proceeds from recognition of an objective reality independent of man's consciousness, and of the knowability of that reality. Recognition of objective reality, which forms part of the content of knowledge, is directly connected with the *concept of reflection*. Knowledge reflects the objects; this means that the subject creates forms of thought that are ultimately determined by the nature, properties and laws of the given object, that is to say, the content of knowledge is objective.[14]

The writer distinguishes this view clearly from the idealist theory of knowledge which "avoids the concept of reflection and attempts to substitute for it such terms as "correspondence", presenting knowledge not as the image of objective reality but as a sign or symbol replacing it."[15] Lenin would furiously protest against the idea that knowledge is a sign or symbol. For him it is a copy of objective reality, a reflection of it, a true image. The writer of the Academy volume cited (who remains anonymous) specifically mentions Ernst Cassirer, the neo-Kantian and his view of concepts as symbolic forms. He continues to argue that, even though modern "knowledge is becoming increasingly symbolical in its expression, and scientific theory often appears in the form of a system of symbols

. . . it is not the symbols themselves that are the result of knowledge, but their ideal meaning whose content is the things, processes, properties and laws studied by the given science."[16]

In other words, the language in which scientific knowledge is expressed may be symbolic, but the knowledge itself is not a symbol, but a reflected image. What then is knowledge itself?

> *Knowledge is the spiritual assimilation of reality essential to practical activity. Theories and concepts are created in the process of this assimilation which has creative aims, actively reflects the phenomena, properties and laws of the objective world and has its real existence in the form of a linguistic system.* [Italics in the original.][17]

In entering a critique of this epistemology, one does not want to be misunderstood. When the Marxist insists that scientific knowledge is objective, he does not deny the subjective pole in all knowledge. He merely insists that objective reality exists independently of our consciousness of it, and that it is reliably, faithfully reflected in our consciousness. His fight is against the subjective idealist who would like to reduce the world to its subjective reflection and to deny the existence of any objective truth. He also fights against the positivist view (e.g., Russell) which reduces the content of knowledge to that which can be objectively proved and verified.

The Marxist would also admit that present scientific knowledge may contain an element of error which will be revealed only by future experience in cognition and practice. In that sense the Marxist is not a positivist, and acknowledges the relative nature of all scientific knowledge. As Lenin put it in his refutation of Bogdanov, citing Engels, and J. Dietzgen:

> . . . for dialectical materialism there is no impassable boundary between relative and absolute truth . . . From the standpoint of modern materialism, i.e., Marxism, the *limits* of approximation of our knowledge to objective absolute truth are historically conditional, but the existence of such truth

is *unconditional* and the fact that we are approaching nearer to it is also unconditional.[18]

Lenin himself cites Hegel's view that Dialectics does contain an element of relativism, but cannot be reduced to relativism. There is nothing static thus in the Marxist theory of knowledge. Marxist epistemology bears striking resemblance to our own Madhvacharya's theory of knowledge, where he regards perception as the flawless (*nirdosha*) contact of the sense organs with their objects. The *Anuvyākhyāna* of Mādhva insists that knowledge gained in perception and validated by the necessary checks yields unqualified reality. Mādhva does not accept the Advaita distinction between different degrees of validity, e.g., between the *vyāvahārika* level and the *pāramārthika* level.

Lenin, at the beginning of our century (1908), faced some of the problems posed by modern physics which today appear crucially relevant to any modern scientific-philosophical epistemology. More than seventy years ago, the questions raised by the British philosophers of science about the validity of empirical knowledge had been raised in a very sophisticated philosophical manner and Lenin was an active protagonist in the debate. Discussing Mach and Lorentz, Poincare and Helmholz, Maxwell and Kelvin, and the general view that in modern atomic physics "matter has disappeared" into mere charges of electricity, Lenin gives a comprehensive survey of the discussion at the beginning of our century about this problem. Marxist epistemology has fully adjusted itself to this problem that the atom can be analyzed as organization of energy impulses. This does not lead to the conclusion that matter does not exist and therefore that materialism collapses. No, Marxist philosophy takes care of this problem today by affirming that it is matter-energy in motion according to the principles of dialectics that constitutes both the world out there and the knowing consciousness. At this point Marxist ontology-epistemology is in the least bit threatened.

Lenin said in 1908:

The electron is to the atom as a full-stop in this book is to the size of a building 200 ft. long, 100 ft. broad and 50 ft. high (Lodge); it moves with a velocity as high as 270,000 km per second; its mass is a function of its velocity; it makes 500 trillion revolutions in a second—all this is much more complicated than the old mechanics; but it is, nevertheless, movement of matter in space and time.[19]

Nature is not a creation of our minds. At this point Lenin leaves us in no doubt. And the modern Marxist goes further to qualify Lenin's copy theory to accept the fact that the reflection of the external world in our minds may not be a flawless image. The modern Marxist is surprisingly willing to accept Karl Popper's theory of science as composed by "conjectures and refutations".

A hypothesis is knowledge based on a supposition. The substantiation and proof of a hypothesis presupposes a search for new facts, the devising of experiments, and analysis of any previous results that have been obtained. Sometimes several hypotheses that are 'tested' by various means are advanced to explain one and the same process. Such elements as simplicity and economy, which serve a supplementary means of determining the most authentic theoretical system, are also of importance in choosing a hypothesis . . .Theory is not something absolute, it is a relatively complete system of knowledge that changes in the course of its development. A theory is changed by adding to it new facts and the concepts that express them, and by verifying (sic) principles. A time comes, however, when a contradiction is discovered in the framework of the existing principles. This crucial movement can be detected by concrete analysis. Its arrival heralds the transition to a new theory with different or more exact principles.[20]

Now that is an admirable summary of the general conclusion of the British Symposium on *Criticism and The Growth of Knowledge*, incorporating the views of Popper, Kuhn and Lakatos, with a slight leaning towards the anti-Communist Popper over against the more liberal Kuhn.

Of course, Popper and Kuhn would be quite innocent of Marxist theory of socially organized labor as an epistemological category. But it is interesting to note that Marxist epistemology has come so close to the Anglo-Saxon pragmatist-analytic philosophical view.

What we have to say in criticism of Marxist epistemology would apply therefore equally to Western philosophies of science, whether of the English-speaking or of the German variety.

It seems to the present writer that Marxist epistemology takes it for granted that our ordinarily perceived reality, purified by theoretical catharsis, negated in its given state, "objectified" in accordance with scientific laws and reconstituted through socially organized labor, is all the reality there can be. To quote Academician S. T. Melukhin:

> The consistent materialist world-outlook has always postulated that the whole world around us consists of moving matter in its manifold forms, eternal in time, infinite in space and is in constant law-governed self-development.[21]

If this position is to be totally consistent, it has to be based on some indubitable proposition and built up from it by clear and consistent methods of argument such as Descartes attempted in his *Discourse on Method*. It seems to me that the Marxist philosophical system is based on two propositions that they have taken as indubitable, but which turn out to be just as problematic as Descartes' *Cogito* though, however, the Marxist system is built up in a neater, larger and more commodious way than that of Descartes.

The indubitable propositions on which the Marxist dialectical system is built up are the following:

(a)   Matter-energy in motion, developing according to the laws of dialectics, is all that exists; this existent reality is eternal, infinite, and self-existent.

(b)   This sole existent reality includes the phenomenon of man who cannot only know that reality as it is given but also changes it in a historically destined direction.

Granted these two propositions, the system develops with

an architechtonic beauty, coherence and comprehensiveness which far exceeds these qualities in Descartes or in any other modern system. Its particular value is that it unites matter, nature, man, society, politics and economics all in one single unifying system of thought. It has no peer in this regard.

If the Marxist philosophical paradigm has weaknesses, they lie at the level of the two basic assumptions, rather than on the methodology by which the architechtonic is built up, though on this latter point there have been and still are impassioned and furious debates within Marxist circles.

In its fundamental assumptions, Marxism has made one fundamental change recently. Previously motion has been seen as a property of matter; today mass and energy are seen as interdependent and interchangeable properties of matter. Einstein provided the two formulae for relating mass and energy,[22] and today the Marxist dialectic does not insist on matter as the only existent; rather it prefers to say "matter-energy" united as one, but with units interconnected by relations of motion, interaction, and structure or system.

The Marxist would claim that matter is infinite and external, though our knowledge of it is finite. They know, as every person knows, that the range of our present knowledge of matter is limited to 10^{-14} cm to 10^{28} cm. The upper limit of 10^{22} cm works out to about 13,000 million light years. Now this is a prodigious range, but certainly the spectrum of reality extends infinitely far beyond ("infinitely is accepted by the Marxists, but not by the present writer, who is a Christian).

To pass a judgment about a reality which one knows to be infinite, based on partial and finite knowledge, would always be hazardous. This is the hazard in the Marxist position. This is also the hazard in the Christian position, for the latter also has only very partial and finite knowledge and yet dares to make judgments about the Infinite God.

The Christian knowingly takes this risk in his faith in God. Without risk there is no faith. But he does not claim that his faith is scientific. The Marxist also takes a similar risk in affirming the infinity and eternity of matter and its ultimate knowability by a finite and mortal mankind; yet he claims such knowledge to be scientific. Let us not make any unfair accusations. The Marxist knows the risk he is taking. In the work frequently cited in this chapter, *Philosophy in the USSR*, Academician S.T. Melukhin makes the following four sets of statements:

1.   The consistent materialistic world-outlook has always postulated that the whole world around us consists of moving matter in its manifold forms, eternal in time, infinite in space, and is in constant law-governed self-development. Nothing in the world exists that is not a certain state of matter, its property, form of motion, a product of its historical development, that is not ultimately conditioned by material causes.[23]

2.   But it is important to remember that matter itself exists only in the shape of *concrete* formations and systems, of which the world possesses an infinite variety. Matter does not exist in "general"; there is no "matter as such outside any definite concrete form.[24]

3.   There are no external causes of the existence of matter; it is the cause of itself or, to be more exact, the concept of cause is not applicable to the existence of the material word as a whole. Its chains of cause and effect are infinite in space and time.[25]

4.   But we must always remember that we know by no means all the universal properties and laws of existence of matter; in fact we probably know only a small fraction of them. After all, matter is infinite, every given system may be an element of a bigger one, any process a fragment of a greater cycle of change.[26]

In the last quotation appears the basic weakness of the Marxist architechtonic. The universe as we know it, says the Christian paradigm (the original Patristic, Eastern, classical paradigm) is a subsystem within a larger system about which we have no conceptual grasp or as yet not clear experimental evidence that can be publicly demonstrated. That larger system extends beyond the present range of scientific knowledge—*i.e.*, from 10^{-14} cm to 10^{28} cm. That system is not in principle unknowable. Many people have known about it, bet their lives on it, and found unquestionable certainty in their convictions about it. But our present scientific methods have not yet been adapted to the knowledge of it. The contention of this book is precisely that science must advance in that direction, as far as it can go.

The universe, as a sub-system with the larger system, is itself regarded by the Marxist as infinite and therefore finally and exhaustively unknowable. He also does not at first accept any larger system beyond the universe open to our senses. We do know some finite limits in our universe—like the speed of light in a vacuum. All known parts of the universe have a finite span of time during which it can exist. The distance between any two given objects is also finite, though it may possibly be infinitely increasing.

The Marxist would admit that his infinite universe cannot have any common absolute time,[27] "no unified quantitative laws of genetic determinacy, no connection between past and future, which are present in all concrete systems", to quote academician Melukhin. Melukhin admits that even the three laws of the dialectic operate only within a finite range:

> In infinity the content of nearly all our concepts and laws undergoes a qualitative change. We are immediately confronted with restrictions on the use of the concept of "system". The infinite universe or all matter may be treated as an infinite number of different interacting objects and systems only insofar as the objective laws of existence permit it to do so. Every system interacts with its near and far environment if its lifetime and distance allow this.[28]

Here is a major admission—that whatever laws we may formulate as "universal", they are unable to explain the whole "infinite system". Marxism's weakness lies precisely at this point—in its conceptual optimism about the universality of its science, an optimism which has little ground in science itself. If the universe is a system, this means that its parts are interacting; according to good Marxist theory today, such actions and interactions cannot exceed the speed of light. If this is so, then space cannot be infinite, for actions in one corner of it cannot extend to the farthest "limit" (which does not exist in infinity) or come back. A system needs interaction, and if the speed of light is the upper limit of action and interaction, this precludes any possibility of an interacting system which is also infinite.

If the universe is a single interacting system,[29] then it cannot be spatially or temporally infinite, given the condition that all actions and interactions can operate only the propagation of material effects, which are limited by the speed of light. So the only kind of infinity left to the Marxist is the numerical infinity of matter; but if this matter is contained in a spatio-temporally finite universe, how then can it in principle be numerically infinite?

The Christian would say the C or the velocity of light is not a limit, because material-spatial is not the only form of interaction between entities and that thought and love are not bound by C. He would insist that the universe is spatio-temporally infinite, and therefore matter can be neither infinite nor eternal. He would also state, though he cannot demonstrate it to the satisfaction of all, that this world open to our senses is only one facet of the Created Order.

The modern Marxist has seen the problem of the non-infinity of space and time. Once again Academician S.T. Melukhin gives us the latest view:

> Instead of the current [Marxist] phrase: 'matter exists and moves *in* space', it would be more correct to say that some material objects or systems move in the spatial structures of other material systems (the Earth's atmosphere, the solar system, the galaxy, the metagalaxy, etc.). Similarly, instead of the phrase 'matter exists and develops *in* time', one should say that time is the duration and sequence of changes in the state of matter. The measure of the duration of this existence of systems is a definite number of cyclical processes in the sub-structure of the subsystems of which they are consti-

tuted (molecules, atoms, etc.) or of larger systems (Earth, the solar system, the galaxy).[30]

In other words, space and time are neither infinite, nor do they have independent existence except as aspects of the structure of matter. Even the four-dimensional space-time continuum is not an independent entity *inside* which matter exists. Melukhin points out the difficulties of a unified Field or General Theory of Relativity, uniting the gravitational, electro-magnetic and nuclear field into one single law.

In this situation, how can one demonstrate the infinity and eternity of matter? Melukhin is quite frank and open at this point:

> What proof can be given of the infinity of the material world? Obviously there can be no complete and final proof because of the very nature of the problem and man's limited possibilities at every future stage of the development of science. Nonetheless, even today there are arguments which suggest that the idea of infinity is not purely axiomatic or postulatory.[31]

Melukhin very guardedly admits the difficulty of *demonstrating* the infinity and eternity of matter. He too has to use the sort of dogmatic language used by the Christian. He moves from the notion of numerical or quantitative infinity to the idea of "structural infinity, which is more or less the possibility of an infinite number of sets of structural relationships. We have already seen quite new sets of structural relationships at the level of what we today call "elementary particles". We are already on the lookout for "quarks", the bricks of which all "particles" are supposedly made. We may have to go down to the scale of 10^{-33} cm,[32] but for the moment we lack the energy to do the breaking down of elementary particles into such micro-micro objects.

The infinity and eternity of matter, as well as its self-existence, can only remain dogmas in the ideology of Marxism, with some corroborative arguments in its favor, which are as old as the pre-Socratics, and which cannot lead to any conclusive atheism. The Christian's position is not basically different. It is acknowledged that some of its fundamental assumptions are not logically or experimentally demonstrable.

Marxism still holds to an ancient Greek classical dogma, the eternity of matter. Anaximander of Miletus (fl 560 BC) had already stated that the Non-limited (*apeiron*) is everlasting, immortal and indestructible.[33] So did Heracleitus of Ephesus, who, *circa* 500 B.C., stated:

> This ordered universe (cosmos) which is the same for all, was not created by any one of the Gods or of mankind, but it was ever and is and shall be ever-living Fire, kindled in measure and quenched in measure.[34]

To sum up, the weakness of the Socialist or Marxist ideology can be located in its unexamined assumptions—such as the infinity, eternity and self-existence of matter, the sophisticated but unsatisfactory epistemology of the reliablity of that which is reflected in the social consciousness of man, and the assumption about the destiny of humanity as a stage in history when society will have no classes or contradictions.

These are dogmatic assumptions. Marxism is partisan in its support of those who adhere to these assumptions and inhospitable to those who question them. Herein lies its inherent weakness.

Much of the propaganda against repression and denial of personal freedom in existing socialist societies has its origin in the Cold War tactics of the market economy world. But there is also genuine repression in Marxism, a great deal of which cannot be justified by the exigencies of the socialist situation surrounded and infiltrated by anti-socialist enemies. Socialism would have commended itself much more readily to all whose interests are not threatened by it, had it been less repressive, both physically and spiritually. One is not talking here of the Stalinist excesses which many socialists would join me in condemning; nor is one speaking about Prague 1968, or any of the decisive Soviet actions against a joint assault from within and without on the world socialist system. One is not even speaking of *Cancer Ward* or *One Day in the Life of Ivan Denisovich*, both of which give exaggerated and one-sided pictures. Neither is one speaking about Roger Garaudy or other revisionists, renagedes and Euro-Communists. One is not speaking about the Maoist critique of super-power obsessions on the part of the USSR. One refuses to be trapped either by the naive anti-Communism of the Americans or the more sophisticated anti-Sovietism of European and Asian socialists.

And yet the ideological weakness persists—the dog-mas which cannot be discussed freely even among those genuinely committed to socialism; specifically the dogma of atheism, the dogma of the eternity and infinity of mat-ter and its self-existence, the dogma of the validity of social reflection as an epistemological method; and the dogmas about the destiny of humanity.

Despite these fundamental weaknesses, Marxist thought is still the closest hypothesis we have in inter-preting current socio-politico-economic reality. Christians have provided no more scientifically convincing interpreta-tion of the current world. We will never be able to arrive at a serious overall paradigm without fully utilizing the insights of Marxist ideology and integrated scientific theory of society.

THE GAMUT OF CHOICES

Looking at the spectrum from the perspective of an Oriental Orthodox Christian, not too well trained in the ways of the West, with some knowledge of the Indian tradi-tion, I would make the following observations:

(a)   The Western liberal tradition appeals to my tem-perament with its breadth and freedom; but it lacks both depth and real content. It is a lazy tradition despite its enormous physical output and achievements. It has still no philosophical system which can serve as the basis for an integrative paradigm. Analytic philosophy is too pedes-trian and averse even to asking the fundamental ques-tions. Existentialism and Phenomenology put too much weight on the individual and the subjective; Structuralism becomes a computer philosophy that seeks to recreate a new idealism which has no way of finding meaning for the whole until the total structural analysis of all reality has been laboriously and exhaustively completed; this is likely never to happen, and many of us will have to pass on from the scene without being able to hear what they have to say about the meaning of the whole.

The liberal tradition's emphasis on critical rationality is something I respect; but I know that critical rationality by itself is incapable of leading me to a paradigm that makes sense of the whole.

(b)   As I look at the Marxist architechtonic, I am impressed with its coherence and beauty, and with the meaning it yields for my action in the socio-politico-economic reality. I need it for my facing that reality. But I cannot accept its epistemology or its ontology, both of which strike me as being rather dogmatic. Nor can I accept its vision of the ultimate destiny of humankind, which also remains at present rather dogmatic in that system. I would like to see the critical rationality of liberalism play a larger and freer role within Marxist reflection. But even Marxism reformed through some application of critical rationality will not yield for me the main contours for an integrative paradigm. I wish that the Western liberal tradition will shed some of its inhibitions and fears and open itself up more to the Marxist ideology. The Western liberal system can only gain from a more fearless exposure to its own weaknesses. But liberalism touched up by Marxism will also not yield the contours of the needed paradigm.

(c)   I could look at the problem first from the perspective of the Indian philosophical heritage—take Nāgājuna, Śankara and Madhva as illustrative of three different options in the Indian tradition.

Critical rationality has much in common with Nāgārjuna though the pyro-technics of logic seems applied to two totally different ends in the West and in Nāgārjuna. Nāgārjuna wants us to move away from pre-occupation with the conceptual, which is incapable of leading us to true enlightenment. In the West, critical rationality leads only to an indifferentism or to too much reliance on individual tastes and preferences.

The Nāgārjuna position is important for me precisely at the point where it coalesces with the position of Śankara of Kaladi[35] and Gregory of Nyssa. All these sages were endowed with great powers of logical reflection such as is rather rare these days. Their power of reasoning was, however, strong enough to make them realize the severe limits of reason in getting to the ultimate realization of truth. Once one has an experience of realization, one can use logic and reason to communicate the meaning of one's experience or to establish its validity.

Gregory of Nyssa acknowledges three faculties of the mind—the world-observing or perceptive, the critical and the speculative. Western thought has in theory given too much prominence to the perceptive or conceptual, and has been more critical than speculative; when it has moved into the speculative, such as in Hegel, it acknowledged no clear criteria or specific tradition and therefore could not sustain any critical development within the speculative tradition established. The monumental character of Hegel's speculative system is witnessed by the fact that it generated two major negative reactions—the Marxian—materialistic and the Kierkegaardian-Existentialist. But the Hegelian system itself could not withstand the onslaught of critical rationality. Far worse, the failure of Hegelian idealism has also led to the discouragement of all forms of speculation in the West.

The Nāgārjuna tradition in India is also anti-speculative; but its use of critical rationality to destroy all affirmations is only a means to leading people to a trans-conceptual enlightenment such as that of Buddha. The Western Enlightenment is an affirmation of rationality and a denial of tradition; the Eastern Enlightenment (Buddha) is an *experience* that *reveals* the futility of every conceptual attempt to grasp the truth.

Nāgārjuna *śūnyatāvāda* or "doctrine of nothingness" is the opposite of the nihilism that results from the critical rationality of the West. By claiming that reality is *śūnyata* or the "void", Nāgārjuna posits, on the basis of an experience, that reality can no more be grasped by the concept than water can be contained in a fishing net. In Asia Zen Buddhism still continues this tradition in a living way—that getting rid of the conceptual and enlightenment or illumination go together. In India J. Krishnamurthi insists that all illusion and all conflict arise from thought, concept and reason; and that ceasing from all conceptual thinking will put an end to all dualism and all desire, bringing genuine peace and enlightenment to all.

If one follows the Nāgārjuna tradition in India, an overall paradigm that holds scientific and other knowledge together in one single framework becomes totally pointless. From my recent hour-long conversation with J. Krishnamurthi, I got the impression that a true Buddhist or Krishnamurthi disciple would also abhor the idea of an overall paradigm as another step in the wrong direction, reinforcing the illusions created by conceptual thinking.

Most likely the Śankara tradition would eventually come to the same position—that energy spent on conceptual clarification of reality through science or through an integrative paradigm would be energy wasted on analysis of the *vyāvahārika* level of reality, an analysis which does not lead to the true unmediated experience which releases one from the bondage to *avidya* or nescience.

(d)   It is here that my debt to Gregory of Nyssa becomes most obvious. He has a very relaxed view of the conceptual and the transconceptual which I find very congenial. In both epistemology and fundamental ontology, Gregory of Nyssa provides us with categories that are still useful to Christians in constructing paradigms.

Human beings are endowed with *epinoia* or the faculty of conception, perception, imagination and critical evaluation. Every art and every science, according to Gregory of Nyssa, is a product of this faculty. He specifically mentions geometry, arithmetic, the physical sciences, technology, navigation, the art of making clocks, as well as ontology, as creation of the faculty of *epinoia*.[36]

Have not all these benefits to human life been achieved by *epinoia*? For, according to my account of it, *epinoia* is the method by which we discover things that are unknown (*ephodos heuretikē tōn agnooumenōn*) going on to further discoveries by means of what adjoins to and follows from our first perception with regard to the thing studied.[37]

Gregory of Nyssa is different from both Śankara and Nāgārjuna in accepting the full validity and usefulness of the scientific enterprise; nay, he insists that science and art are both from God. "Thus human life invented the Art of Healing, but nevertheless he would be right who should assert that Art to be a gift from God."

But Gregory makes also the point that precisely because this gift is exercised in freedom, "no one would deny that he who has learned to practice an art for right purposes can also abuse it for wrong ones, so we may say that the faculty of thought and conception was implanted by God in human nature for good, but with those who abuse it as an instrument of discovery, it frequently becomes the hand-maid of pernicious inventions. This potential double use, for good or evil, applies to all our faculties, according to Gregory. The fact that science has been misused is a witness says Gregory, to the opposite fact that it can be used for good purpose.

This is a slightly different position from that of Mādhva, for whom the concept of *sakśin* (witness) is in some ways similar to that of Gregory's *epinoia* (conception). It is this inner witness within each of us that perceives sense-knowledge as well as other objects which cannot be perceived by the senses (abstract thought, imagination, speculation). It is this same inner witness within consciousness that also experiences certainty in knowledge.

The similarity and difference between Nāgārjuna and Śankara on the one hand, and Mādhva or Gregory on the other, can be put thus: they are all preoccupied with an ultimate experience of self-realization or of becoming what one is. But Śankara and Nāgārjuna have a lower evaluation of the conceptual than Mādhva or Gregory. For the first two the conceptual is something to be *overcome*; for the latter two it is something to be transcended. None of the four would hold to the view often current in the West that truth is propositional. Truth is that which is, rather than that which is stated.

Gregory of Nyssa states clearly that concepts are human creations, that they can be good or bad, right or wrong, but also that words and concepts do not constitute truth. It is possible to call light darkness and darkness light (Isaiah 5:20); but Gregory says that that is a form of drunkenness. Science is made up of perceptions and conceptions based on theories or hypotheses which we have created as human beings.

These words, concepts, theories, hypotheses, which constitute science are necessary for the full growth of man, but full growth requires that he go beyond these. Śankara, Nāgārjuna and Krishnamurthi would seem to deny any point at all to these conceptual formulations, whereas Mādhva and Gregory would think of them as legitimate and necessary processes through which humanity has to advance to something beyond.

Our paradigm must then do justice to science and yet leave us free to move beyond.

CHAPTER ELEVEN

Science and Meaning

Our modern scientific effort is, after all, only a few centuries old. Modern science is still young and vigorous. It should and could make some new breakthroughs. Centuries old habits, which have yielded magnificent results in the past, may have to be abandoned in the process.

There are two basic directions now seen where we can expect some significant breakthroughs—one in Western liberal scientific thought, and the other in Marxist dialectical humanistic scientific thought.

In the West, Abraham Maslow is the pioneer and fertile genius of a myriad ideas, none of which has as yet become fully recognized or accepted by the scientific community. In his 1966 work on *The Psychology of Science*[1] he makes an interesting distinction between mechanistic science and humanistic science.

Professor Maslow saw clearly the damage that a mechanistic approach to the human psyche was doing to the infant science of psychology. He charged that "the mainstream methodology in psychological research, mod-

159

elled after the mechanomorphic tradition of the physical sciences, veils us from a fuller knowledge of human personality—a knowledge which we sorely need."[2] Traditional psychology, Maslow accused, was misinterpreting human personality as solely composed of features which we can measure or manipulate.

Now the problem raised by Maslow goes far beyond the discipline of psychology. It is an accusation against mainstream science itself, which creates its own *Weltanschauung* and subculture, which in turn affects the perspectives of society as a whole. As Maslow puts it in the preface:

> In the broad sense, Science can be defined as powerful and inclusive enough to reclaim many of the cognitive problems from which it has had to abdicate because of its hidden but fatal weakness—its inability to deal impersonally with the personal, with the problems of value, of individuality, of consciousness, of beauty, of transcendence, of ethics.[3]

Scientific experience itself has exposed the weaknesses of the mechanistic and fragmentary approach, and has pushed us on to the study of the human and the holistic. But this new science remains essentially underground because of an oppressive ethos in the scientific community. Maslow's alternative was certainly not to freak out from the rigorous demands of experimental science and pursue a nebulous Zen or Tao with an unhealthy emphasis on personal experience over against the experimental; nor did he stand for an impulsive whimsicality that becomes easy to laugh at or scorn.

The psychology of science seems just as important as the philosophy of science and the two should be considered integral to each other, despite all the strictures about psychologism in philosophy or epistemology. Science itself needs psychoanalysis; it is somewhat sick. Science has attained much by being careful and cautious, slow and patient, averse to accepting hearsay as evidence, insisting on clear and indubitable demonstration of everything before its being accepted as true. This creates a particular kind of personality—one that is a bit overcautious

and smug at the same time, refusing to look except where
its own limited light falls. It mechanizes and dehuman-
izes the scientific observer himself, as well as the reality
he observes. To try to avoid the subjective, or to think
that the subjective can be avoided in our dealing with
reality, is a sort of sickness—one that has been useful,
nevertheless.

This neutral-objective stance has been a false pose.
It has distorted not only the perceived reality, but also the
perceiver. Unscientific assumptions abound in this false
pose and this self-deceiving commitment to "objective
truth." Strict causality was once assumed as an axiom;
the man-made machine was once taken as a model for all
reality. Today we know that these are false assumptions;
but we still operate too often on the basis of these. Abstrac-
tion and reduction, without which science can hardly exist, are too naively accepted as self-evidently valid.

A few of our contemporaries like Abraham Maslow
and Michael Polanyi tried to tell us that personal knowledge is radically different from nomothetic, axiological,
"objective" knowledge of things. But we are still finding
it difficult to overcome the ingrained habits of two centuries of training, to explore the possibility of extending
science beyond the measurable and nomothetic. The scientific community still has an ethos that is oppressively an-
tagonistic to ways of thinking like that of Polanyi and
Maslow. This ethos is loaded with a deep sense of insecurity and this is where the sickness lies—in science's inability
to recognize its own basic insecurity. Maslow calls it the
Fear of Knowing or Fear of Personal and Social Truth,[4]
which leads to a grand resistance even to attempts at
knowing the truth.

We still have difficulty in recognizing that just as
there are only three realities—self, world and God—there
are only three ways of knowing, which are interrelated but
different. The way I know my own self and derivatively
know other selves is basically different from the way I
know things. And the way of "knowing God" is unique.
Why should science be limited just to the knowing of the
world? Why should we insist that the way of knowing the

world is also the way to know the self and God? Why cannot science seek to find and develop reliable ways of knowing and dealing with persons, as well as with the meaning-structures of existence, i.e. the various ways in which people have found meaning in the past and discovered how to benefit from this knowledge for our own dealing with reality in the present day?

These seem to be the areas where science should seek a new breakthrough. The breakthrough type of research still seems to work underground because of an unnecessarily oppressive ethos in the scientific community, which that community finds difficult to acknowledge.

Of course one does not want to be ungrateful to the many who have made positive contributions to the knowledge of persons and of meaning even within the framework of present science. I can think of the enormous wealth of cultural and ethnological information gathered, the experimental research of, say, a Jean Piaget in child psychology and so on.

But we have many fields of research which remain still basically disrespectable or unrespected, at least unrecognized and unappreciated—dream research, brain function research, research in biofeedback training, altered states of consciousness, communication with plant life, Kirlian photography, studies on fire-walking, meditation, breathing suspension, psychic healing, paranormal phenomena and so on.

Of course, it is quite difficult to separate the wheat from the chaff in all the literature that comes up now in all these fields as well as in the occult, in astrology, magic, U.F.O.'s and all the rest. But are we sure that the wheat therein is negligible? Perhaps as we remove the chaff, we may find more than wheat, perhaps precious gold and diamonds. Scientists themselves should read some of this literature before they reject it as total nonsense.

The breakthrough we look for in science is, then, the way to develop methodologies for gaining reliable and useful knowledge on the interpersonal, the social and the transcendent without being bound by the methodologies of the "lower" or physical sciences.

The socialist countries seem to be less insecure than the liberal West about the reliability of science. They have more fearlessly entered into research on several of these "underground science" topics like psychic healing, Kirlian photography and supersensory perception.[5] When they see an unusual phenomenon like a psychic healer or a halo, they do not write it off as a freak occurrence, for fear that it will upset their general theories. On the contrary, precisely because of their confidence that reality is one and mutually coherent, they pick up freak instances as indicators of a realm of truth which is hid from our normal scientific perception. In pursuing the freak phenomenon they assume that whatever emerges as reliable knowledge in that sector can only enhance and improve the quality of the sum total of our scientific knowledge—though when it comes to the matter of meaning structures they may also be inhibited and insecure.

The Socialist countries insist, however, that "the final object of all cognition is objective reality"; and add that there is neither complete continuity nor total discontinuity between common sense knowledge and the various forms of scientific cognition. Science and common sense are in dialectic relationship; both are in the process of dialectical transformation.

Furthermore, the Socialists of Eastern Europe firmly believe that Man is the sole creator of meaning or value or sense in this world. In a recent article in the Polish philosophical quarterly *Dialectics and Humanism* (Spring 1979), Professor Janusz Kuczynski poses this question: "Does Being, conceived as a whole or the objective reality, as the sum of what was and is, have any sense?"[6]

The Marxist answer to the question is: "Being acquires sense for man when he becomes aware of its structures, its potentialities for development, its wholeness." I agree with the Marxist at this point, and I accuse Western liberalism of dismal failure at this point of holistic awareness.

My disagreement with Marxism begins at the point of analysis of how this "awareness of reality's structure and potentialities" is to be undertaken. I can gladly and enthusiastically agree with my friend Professor Kuczynski that this awareness cannot be wholly propositional, and that meaning and sense can also be expressed in symbols. In fact certain symbols are more powerful than any conceptual understanding to inspire action. History itself can be kept in our awareness much better by ancient buildings carefully preserved, by monuments and artifacts than by written texts.,

But all meaning or sense is relationship. Relationship to what? To objective reality, past and present, says the Marxist. The Victoria Memorial in Calcutta becomes a symbol of colonial oppression only when one realizes not only that it was built with the sweat-labor of the Indian people whom the British were exploiting, but also that it was built to humiliate the Indians who had built the Taj Mahal. Unless one sees cultural history a bit theoretically in terms of the ambitions of the colonial masters to subdue India's millions by outdoing their architectural masterpiece, the Victorial Memorial remains a large ugly building that carries no symbolic sense. An Englishman who seeks to justify British imperialism in India would see it in other terms; it could be for him a symbol of the great and glorious times of the British Raj, without any reference to the Taj, which it was supposed to outshine.

Thus, not only is man the creator of meaning or sense even in the use of symbols, but the very meaning-structure he creates is heavily influenced by his own interests and perceptions; this leads to the same symbol having different and contradictory meanings, depending on where one stands and what one stands for, in the dynamic process of history. We also see how symbolic and propositional truth are intertwined, how they are complementary.

This relationship between the sense-giver and the meaning-object (the signifier, the signifying and the signified, if one thinks in structuralist terms) should become the object of scientific study; but a study of this relationship is, of course, dependent on our understanding of the meaning-giving subject and the meaning-giving object.

There are dozens of layers in both the subject and the object that need to be analyzed, in a manner that goes beyond a simplistic structuralist semiology. We must take seriously the Marxist charge that Augustinian or Thomistic or even Teilhardian ontology is also too simplistic—in attributing to God all three functions; that God is the source of all meaning, that He is both the subject and object of all meaning, and that He is also the relationship. The consequence, as Kuczynski points out, of this simplistic Christian position is that man only *discovers* but does not *create* meaning; that meaning is not relational but absolutely objective and given; that man is merely a passive perceiver and appropriator of meaning. Perhaps the Marxist caricatures the Western Christian position when he says:

> Thus God seems to mark out a limited ontological horizon and, by the same token, a clearly defined horizon of sense—from the point Alpha to the point Omega. It is of no importance in this case that the idea clearly points to a "spatial" and temporal end of all human endeavor. The important point is that human activity thus proves to be demarcated a priori. The human being ceases to function as a creator while his role is limited to that of a mere executor of plans devised by God, the Creator.[7]

The Polish Professor admits that in practice this limitation may have no negative influence on Christian activity. But the point he makes is worth heeding: "Nevertheless any variety of the Christian religion and any religion where God is so unlike man circumscribes the ontological and the temporal horizon."

The kind savant from the Polish Academy of Sciences even exempts "those varieties of Christianity which try to overcome the pitfall of the 'opium of the masses'—like the truly evangelical, Teilhardian and the communist (sic) Christianity"—from these negative practical consequences. And I believe that in his term "truly evangelical" he would include some forms of Eastern Christianity. Of course Professor Kuczynski's acquaintance with classical Eastern Patristics is understandably limited by his Polish background. Otherwise, he would have realized that the point of Eastern Classical Christianity is precisely the basic *similarity* between God and man—so much so that it is the Son of Man who now sits at the right hand of God and runs the universe. There is no idea of the "immutability of human nature" in Eastern Christianity. Neither does the Eastern Tradition, which emphasizes the freedom of humanity as its constitutive element, ascribe to God such sovereignty that the actions of humanity make no difference to God's plans. God does not create "meaning" in the aprioristic sense which Kuczynski attributes to Christian thought. Man is co-creator, not only of meaning, but also of reality itself. This is axiomatic in Eastern Patristic philosophy.

Professor Kuczynski is a Marxist who is very kind to Christians and recognizes the value of a deep level Marxist-Christian dialogue: "Let me emphasize it with full force: the contribution of the Christians to the creation of the sense (meaning) of Being and history has recently increased considerably."[8]

He thinks that the problem with Christian thought is our aprioristic ontologism; that our understanding of the meaning of Being is given a priori and not creatively formulated by persons in socio-economic evolution.

I would not agree with this statement about the Christian approach to meaning. Neither would I accept Professor Kuczynski's position that "the sense of Being depends, genetically speaking, entirely on man, the unique creator of sense," though I would agree with him that it is in the process of actively changing reality that humanity creates sense. I do not agree with him that all creativity is necessarily individual. Nor do I think that this is an essential axiom in Marxism. The individual's creativity can only be one manifestation of human society's creativity, and cannot be conceived independent of that society which gave birth to and shaped that individual.

The Marxist contention is that both values and meaning are human creations, and nothing more than human creations. Humanity, according to them, is the demiurge of meaning, and through that meaning-creation, the demiurge of reality itself.

The Christian would say that all meaning, as far as

we know, is perceived and appropriated by human beings, and that human beings can to a certain extent shape reality on the basis of their grasping or not grasping meaning. But we would also affirm that precisely in the process of grasping that meaning and transforming reality, man becomes aware of the loving and wise Power from whom comes not only both the self and the world but also the meaning itself. The Christian realizes that the language and categories or symbols in which man expresses the meaning of being are human creations. He would also insist that such human expression, whether in propositions or in symbols, can never be exhaustive; he would even admit that there may be basic differences within the Christian community itself, in the enunciation of that meaning, and therefore of the ways in which Christians seek to shape reality. But it is not possible for the Christian to argue either that man is the sole creator of meaning or that the meaning is so objectively given as to preclude all differences in the articulation of that meaning.

We come back to Professor Kuczynski's article as one of the clearest pieces of recent Marxist writing on the subject. Let us quote his own words:

> It is clear then that the sense of Being cannot be anything that is discovered or granted once for all, but rather that it must be something which is dialectically developed through evolutional, quantitative growth, through scientific revolutions, through the drama of human cognition, and primarily through man's practical attitude towards the universe.

The Marxist scientific interest in the meaning of Being is one that others should take note of. It is this quest that Western science and philosophy have practically abandoned. One of the major pleas of this book is that Christians should pick up the challenge of integrating science and philosophy in a pattern that gives meaning to Being and orientation to existence. This indeed is what is meant by Christian humanism. We too should have the strength to draw the general conclusions (tentatively) from our scientific knowledge, historical experience and the philosophical traditions of humanity, in order to create a meaning-giving pattern. Here the breakthrough is possible only if scientists, philosophers, historians and theologians can get together and work together in a disciplined manner, in a sort of Christian academy where such integrative thinking can be undertaken in a systematic way. We could begin by making a survey of how humanity has sought to find meaning for Being in the past; proceed then to contemporary efforts, implicit or explicit, to find meaning for the whole; enter into dialogue between various cultures and ideologies on the same point.

The breakthrough will come, at first imperceptibly, but soon more clearly, only through a trans-academic community that pioneers in new patterns of living and worshipping together and producing things and thoughts together and in that process shaping each other. The Christian academy would probably be something that could be started right away if the money and the personnel were forthcoming, but the development of a systematic pattern, or patterns, of meaning is more likely to emerge in actual communities where men and women can pioneer together to create a new style of life and a new way of living for others. The task of the academy would be to generate such trans-academic communities and then to have their experience with life and its meaning fed back into the academic community, for purposes of more systematic reflection and feedback into the experimental communities.

The breakthrough in science will not come from one such academy with feeder trans-academic communities alone. It will be a many-pronged and largely uncoordinated effort, occurring in various sectors of world society. The important thing is to put the search for meaning at the heart of the academic enterprise, to make sure that science, technology, schools, universities and research centers, history, philosophy and political economics all give a central place to the search for the meaning of Being.

Side by side, underground science will flourish. The underground science will seek at certain points to be holistic, as it now does, but without sufficient sensitivity to the diversity within the whole. For example, the Association for Humanist Psychology is at present absolutely naive on matters of political economics or ideology. In the Soviet Union, where political economics is largely taken for granted, the new research into the paranormal remains unintegrated with the general perception of reality in Marxist ideology. The Christian Church, with its conception of Catholicity as concern for the whole (*kata-holikē*) cannot run away from this task of holistic integration of knowledge and its transmutation into wisdom and love and power, in unity, for the welfare of the whole.

As Christians, we disagree with the Marxists in their assertion that the creation of a meaningful unity is a purely human task. We are convinced that the Holy Spirit is at work, drawing all things into unity in Christ, a unity to be made manifest beyond history, but one that seeks less and less imperfect manifestations within history. We do not even dare to conceive all the contours of this final unity; but we seek partial perceptions of the whole, and partial transformations of the whole in the direction of its final perfection.

The point is that science at present has not accepted its full responsibility in the carrying out of this task. Marxism is better off than Christianity at this point. The Marxist recognizes that the process of development is a gradual and dialectic process of personal and community development, a differentiated and integrated actualization of generic human powers and values achieved by interaction of humanity with the world through organized social labor.[9] This process is to be advanced through science and technology, applied within a socialist political economy. Such a development builds the substructure for the basic needs of human existence in order that humanity may pursue the higher values of culture and meaning—"such as truth, beauty, autonomy, friendship, love, justice, and the like," as Professor Parsons, citing Professor Maslow, puts it.

Marxist humanism at its best is committed to the utilization of science and technology for fulfilling the higher nature of man. Christians often caricature the Marxist approach as "materialist" without realizing that most Marxists are more humanistic and less materialistic than most Christians. Christians have as yet to develop anything like an integral humanism to meet the complexity and sophistication of the Marxist perspective.

Finally, Christians have yet to overcome the tendency to look at science as an enemy or a rival. But then neither do they need to be hypnotized by the achievements of science into believing that it is all-powerful. Science is a human creation that we can use for refurbishing our social and material as well as our meaning-related existence. In order that science may serve us in all three areas —as the material base of existence, in the socio-cultural shaping of it, and in moulding it for meaning-perception and meaning-related existence, science needs to make several breakthroughs.

CHAPTER TWELVE

Science, Technology and Philosophy

SCIENCE, PHILOSOPHY AND THEOLOGY

In the title of this section we have reversed the order laid down by Auguste Comte in *La Philosophie Positive*. For Comte the human intelligence starts with theological theory, moves on to a metaphysical theory, and arrives finally at a positive scientific statement of the matter to be known. For him this was a great natural law he discovered in the totality of the progress of human knowledge from its inception to our time. Each branch of human knowledge, according to him, begins in theology to end up in science. The metaphysical is only a transitional stage.

Comte did not say theology, philosophy and science; he spoke of three stages of theoretical formulation in each branch of knowledge:

(a)   the theological or fiction stage
(*l'état theologique ou fictif*)

(b)   the metaphysical or abstract stage
(*l'état metaphysique ou abstrait*) and

(c)   the scientific or positive stage
(*l'état scientifique ou positif*).[1]

We are not proposing now the three stages in reverse —first science, then philosophy and at last theology. We are arguing for the legitimacy of all three forms and others, of apprehending truth and holding on to reality. We do not even disagree with Comte's delineation of the grand lines of European intellectual development so far.

We can also today agree with Comte's view, already expressed a hundred years ago with great foresight, that as science grows and perfects itself, the other two disciplines, the theological and the metaphysical, which then coexisted with the scientific, would recede in importance and acceptance. This is what has actually happened in European or Western intellectual life.

Comte also argued for a "homogeneous doctrine to be formed by combining all the generalities of positive knowledge in the various branches of knowledge. He thought, however, that this would take place spontaneously. He did not suggest one Single Grand Law that would explain everything. The unity he sought was rather in Method. As for laws, while the lessening of the number of laws was to be desired, their unification was unnecessary and impractical.[2]

Recognizing the partial validity of Comte's grand vision of the history of European thought, it is time to ask whether the process should go on as it has in the past. Has the time not come to retrace our steps, and to relate the deliverances of the scientific method in various disciplines, to the disciplines of theology and philosophy, or to philosophy in the larger sense, both religious and secular? Retracing is only to get orientation for the way forward.

THE CRITERION OF MODERNITY

We often speak of "modern science" and "modern philosophy" without stopping to think what "modern" means. Does it mean simply "recent or "current"? It certainly should not. What then is the criterion of modernity?

In seeking the answer to that question we come up against certain presuppositions of European history and intellectual life. We can probably say that modern philosophy begins with Immanuel Kant and René Descartes. It would be more difficult to fix the origin of modern science with Francis Bacon or Isaac Newton.

Mathematical physics probably came of age in the 17th century through the work of Descartes (1596-1650), Galileo (1564-1642), Kepler (1571-1630), and Newton (1642-1727), even before the experimental method became normative with F. Bacon (1561-1626) and J. Locke (1632-1704).

Characteristic of Descartes was the rejection of authority—the authority of the Church and the authority of Aristotle. But while rejecting Aristotle's *conclusions*, Descartes followed his *method* of syllogistic reasoning. Even Descartes had, however, to start with some clear and indubitable, self-evident first principles from which the reasoning could start and by which it would be guided. These first principles could not be arrived at by syllogistic reasoning, but only by inductive insight, or "the natural light of reason".

Descartes was only half modern. He could appreciate Galileo's use of mathematics and mathematical reasoning. But he despised the empirical approach of Galileo which he called "continual digressions".[3] Francis Bacon on the other hand rejects rationalism and extols empirical observation and experimentation:

> There are and can be only two ways of searching into and discovering truth. The one flies from the senses and particulars to the most general axioms, and from these principles, the truth of which it takes for settled and immovable, proceeds to judgment and to the discovery of middle axioms. And this way is now in fashion. The other derives axioms from the senses and particulars, rising by a gradual and unbroken ascent, so that it arrives at the most general axioms last of all. This is the same way, but as yet untried.[4]

The method which Bacon denounces as just fashionable is the revived platonism and deductive rationalism of Descartes. Bacon prefers the inductive method, but seems innocent of its philosophical problems. Can we say then the following criteria are decisive for "modern thought in philosophy and science?

(1) Rejection of the authority of tradition, and its replacement by reason.

(2) Commitment to the empirical method of knowing by induction.

Descartes and Kant accept the first principle, but not the second. Modern science accepted both principles, but has found both problematic.

SCIENCE, PHILOSOPHY AND METHOD

Science cannot reject the authority of tradition. It has certainly done without the authority of religion. But modern science itself is a tradition, cumulative all the time, with occasional revolution, a continuous process where each generation appropriates and builds on the work of the previous generation. Of course each generation innovates, rejects some, renews some, and adds some. But that is the true nature of all traditions— including the religious traditions. Science, whether ancient or modern, lived on and lives on, only by the force of tradition. As philosophy through Habermas and Gadamer has clearly shown, there is no other way.

On the question of the empirical-inductive method of science, it is philosophy again that has showed us that while this method is operationally successful, it does not yield any logical certainties. It is absolutely necessary for human beings to know this limitation of the scientific method. It was in the absence of such philosophical knowledge that scientists and philosophers made pretentious claims about the potential omniscience and omnicompetence of modern science. Were it not for philosophy, science could have seriously led us astray.

Of course, philosophy itself made even more preposterous claims to its own all-knowing capacity. It was modern science again which showed philosophy its own limitations. Philosophy's platonic disdain for the particular and the detail has now been sufficiently laughed at. Philosophy has now bowed to science as the queen of the académe and is anxious to conform to the latter's reign and rule, seldom daring to dissent or revolt.

In trying to be "modern", rejecting authority and tradition, philosophy has lost its grip on itself. Very few schools of philosophy survive for many generations. Nor does any grow by simple accumulation or revolution. Each philosophy may leave a little residue that lasts longer, but philosophies are simply replaced, not renewed or reformulated.

Consequently philosophy lives in an ethos of demoralization. Our best minds prefer to get into science rather than philosophy. In terms of number of persons and volume of resources devoted to the two disciplines, there is absolutely no comparison. Auguste Comte has been proved right at that point. The only survival from Natural Philosophy has been mathematics, or mathematical logic. Comte actually proposed an education program—"the rational plan of the study of Positive Philosophy. . . MATHEMATICS, ASTRONOMY, PHYSICS, CHEMISTRY, PHYSIOLOGY, SOCIAL PHYSICS".[5]

It is to be noted that even the Positivist Comte, while rejecting metaphysics as representing the transition stage from which we have evolved, did not discard philosophy as such—particularly the mathematical science. Cybernetics and computer technology are simply technologization of the mathematical family of sciences, but they are hardly regarded as philosophy.

What then is Philosophy? "Knowing one's way around" —says Prof. Wilfrid Sellars of Yale.[6] It is having one's "eye on the whole", without staring at it all the time, which distinguishes the philosophical enterprise, according to this enterprising philosopher. Whether the philosophical community would agree with him is a different matter. Philosophy, he says, cannot be purely analytic. It has to be synthetic, though analysis is a precondition of synthesis. At this point the American comes close to the Marxist view that true philosophy is a higher level of universalization of the deliverances of the distinct sciences. No scientist would today deny (I hope) that there are notions and concepts which we use in science that need analysis and reflection—e.g. sequence, causation, order, relation. If one asks the question why there is order in the universe, immediately one is beyond the realm of physical sciences, into philosophy. Or if you ask the question of absolute origins in the chain of causality, the question about the first cause, we are already beyond the limits of science.

The "eye on the whole that philosophy recommends, where it is not in the exclusively analytic bind, needs to be opened. That is the task of philosophy; but even philosophy is unable to cope with the questions of origin and destiny. The Heideggerian question "Why is there any being rather than no being at all?"[7] can be *discussed* by philosophy, but not *answered* by it, as Heidegger himself found out.

Science needs philosophy at another point. Many creative scientists will tell you that it is philosophy, especially Asian philosophy of one kind or another, that has often inspired them to an alternative vision which later became a new theory or a new discovery.

The fact that Western philosophy seems mostly unable to serve this function for scientists is significant. One reason may be that Western philosophy has already lost its "eye on the whole", or is at least inhibited by science itself from opening its eyes to the whole.

Modern science and modern philosophy as well as the secular frame of reference in which both have developed are children of the Enlightenment, of the revolt against Tradition and Authority and by implication also against the God of tradition and authority.

Today's challenge is not to retrace our steps, but to trace them back and forth, in order to get our bearings again. We must move forward, but we will not get the orientation necessary until we have understood the way we have come. We have to move forward in science, philosophy and theology. I believe that the movement forward, radically changing direction, will take place only in a concerted effort to shift gears in all three—in science, in philosophy, and in theology.

THE HEIDEGGERIAN CRITIQUE OF SCIENCE AND PHILOSOPHY

Martin Heidegger (1889-1976), so little understood in the West, has some incisive things to say about science. His dictum "Science does not think", was then qualified to say "Science does not think in the way thinkers think."

> All the sciences (*Wissenschaften*) have leapt from the womb of philosophy, in a two-fold manner. The sciences came out of philosophy, because they have to part with her. And now that they are so apart, they can never again, by their own power as sciences, make the leap back into the source from whence they have sprung.[8]

Mathematics does not say what mathematics is—by the mathematician's method. The artist does not say by his art what art is. The physical scientist cannot, by the methods of physical science, say what physical sciences are. And according to Heidegger, the modern philosopher is unable to say what philosophy is, by his present methods. For him philosophy requires a different stance from that of science.

Heidegger does not believe in the empirical method or logical reasoning as the mainstay of philosophy. For him, to know what philosophy is, itself requires a change of stance from that of common sense or modern science. Some may want to call him "premodern" but in his own self-understanding he is "postmodern".

Heidegger refers to Aristotle's definition of philosophy as enquiry into the "primordial principles and causes (*prōtai archai kai aitai*) and the cultivated aptitude or competence for correct vision (*epistēmē theōrētikē*). But Heidegger himself wants to go beyond Aristotle to define philosophy as an attuned responsiveness to the Call of Being. Being (*Sein*) appeals to us, not just in our hearts, but through the things that exist (*Seiendes*). To "think is, for him, ultimately to be attuned to the Call of Being, Being which conceals itself behind the existents, and withdraws as we approach. The very withdrawal is a call, the call to follow. And to do that kind of thinking is for Heidegger what it means to be human.

At this point Heidegger comes very close to Gregory of Nyssa. Gregory's understanding of the Call of Being is explained through the myth of the bride (*The Song of Songs* in the Old Testament) running after the Bridegroom whose voice she hears, but with whom she cannot catch up because he continually recedes and hides himself.

This Call of Being does not come to us through science, and not until we change stance. It can be heard in music and art, in our marveling about beauty, in our standing before the entrancing and beckoning vision of truth. It can be heard above all in poetry, which is the most creative use of language—to reveal the *logos* which is concealed.

The Call of Being is a summons to be human—to understand the nature of the questioner who questions Being about *its* nature. The questions which modern science and philosophy have asserted too dogmatically to be unaskable and unanswerable, became the central questions for philosophy—about the ground and origin of the Being that manifests itself through the beings in the world and through oneself.

Here philosophy enters into its most central task— that of going beyond science out of the questions being asked by science, ultimately to find a plethora of answers in various religious traditions—and then finally to go beyond all, science, philosophy and theology, to find . . . to find what? *To find that which cannot be named or described in words*, but can be evoked through art and music, poetry and religious discourse.

That, for which we seek, is not just another being among beings, supreme being or the ground of all being, something that stands over against us. This "that is not being in the sense we understand beings", but a "that which goes beyond all being and nonbeing", that in which our consciousness exists, along with the object of that consciousness, the world.

Being is not a concept or a predicate, one that we can catch with our knowing. In our quest for Being, the ontological enterprise which attempts a *logos* or discourse about *to on*, the Being, can only show us the *Ontic* that stands beyond ontology, behind and underneath all logos or discourse. Discourse does not lead to Being, but only to being-ology to coin a bad neologism.

The gap that looms large between Being and knowing, the abysmal Abyss, is at the root of human alienation, and this is the problem that Heidegger made his central concern.

COMMUNION AS DISALIENATION

The plea of this book is to appeal for a retracing of our steps—back through science, philosophy and theology, reversing the Comtian course of European thought, to what is the alpha and the omega, the source and the destiny.

We will have to come back again, after having drunk at the source, back to theology, philosophy and science, but with the Call of Being ringing in our ears and heart. Then and then only can we find new orientations in living, thinking and acting, as renewed human beings, as a renewed human community. We can then see ourselves and the world in the light of Being and get new orientations on the way of being—the way of living, thinking and acting —the way of love and wisdom and power.

It is only then that we can be, be-with-others, be-with-the-world, in our being-with-Being. To be human, authentically human, is communion—communion with others, communion with the world, rooted in communion with Being. To be authentically human means to be free from alienation, and from the guilt and the anxiety that spring from that alienation. To be authentically human is to be free, free from loneliness and despair, from selfishness and boredom, from injustice and exploitation, from cruelty and oppression.

Disalienation is communion. Science, philosophy and theology, when done in the alienation of aloneness, became further alienating. The reconstitution of all three disciplines demands communion which is the true being of community.

Today the new demand on humanity is to break through the barriers that bar community, community not just across disciplines, but also across global humanity, and with the world we experience, both animate and inanimate.

But that communion has to be more than just an "eye on the whole". It has to go beyond the eye and the mind, to the heart, from which attitudes spring. The heart here does not symbolize feeling. It refers rather to the center of our own being. Mind and concept can also be alienated from that center. In fact that is the more serious and fundamental alienation in humanity today.

Science and the scientific culture bear a large share of responsibility in the alienation between heart and mind, between the perceiving center and the eye through which the center perceives. Here again the desire of Anglo-Saxon science to attain objectivity in thought by ruling out the subjective as the source of error is at fault.

We have learned the art of critical thinking, and the capacity to reflect on reflection itself is a peculiarly human gift and achievement. Yet that reflection on reflection has itself to go further. That going further demands "caring" or in Heideggerian terms *Sorge*.

Language is the tool and the product of thinking. Without language we are unable to reflect, either on experience or on reflection itself. Again to use Heideggerian language, language is the "House of being"—that sphere in which the human person can dwell aright and make clear to himself/herself who he/she is.[9]

And yet language can be the most alienating of all our gifts. Especially when language is used to capture truth and reduce it to propositions and mathematical formulas. Language can be used too exclusively to dominate and possess beings, and thereby obscure Being itself. Language can create the illusion that we have the Truth. Truth cannot be possessed. It cannot become property. Language, by creating the illusion of possessing the truth of beings, can make us deaf to the incessant Call of Being itself—the call to be and not merely to know. Harold Aldermann, in a perceptive essay on "Heidegger's Critique of Science and Technology"[10] lists four ways in which the scientific way of thought and the genuinely philosophical way of thought differ. First, scientific thought adopts a beings-oriented, domineering style, while philosophical thought seeks to be attuned and responsive to Being itself. Second, the scientific language is incapable of helping us understand the nature of science itself which reflects on experience on the basis of logic and mathematics alone. Third, Heidegger claims that science is the culmination and development of Europe's metaphysical quest which fails to disclose the meaning of Being. An alternate method or way has to be found to open our ears to the Call of Being. Fourth, Science and Philosophy have two different intentions and therefore use language differently, one to describe, the other to evoke.

Science meddles in the realm of philosophy by proposing to offer us world-views or world-pictures, which is not its proper realm, according to Heidegger. The attempt to construct a Weltbild and to equate it with the sum total of being, is notoriously popular among some breeds of scientists. This attempt to make an objectively true representational map of reality is doomed to fail, because it is at best a collation of the representations of being available to discursive mind and senses.[11] Descartes' *res cogitans* and *res extensa* are put together as universe or *Kosmos*, in the hope that we can thereby possess the knowledge of the whole truth by the scientific method. Heidegger insists that science has access only to beings, not to Being. Science deceives when it tries to put together the beings coming under its knowledge and to give the impression that the collation of beings constitutes Being. Communion does not result merely from putting things together in thought.

Theology, on the other hand, makes the same mistake. The medieval European Church claimed to have a revelation from God about the hierarchy of Being with God at the top, and then sought to reduce that revelation to doctrine or dogma. This attempt to know and possess revealed truth (as *scientia* in medieval terminology) was doomed to fail and has failed. The truth was assumed to be contained in the two sources, tradition and scripture, and reason simply was to draw out the truth in propositional form. The Church also set itself up as the possessor and arbiter of revealed truth and its propositional formulation as dogma.

It is in the breakdown of this theological *scientia* that modern science has its origin. The medieval theologian also posed questions to be debated about revealed truth (*quaestiones disputae*) and by marshalling the (logical) arguments pro and con, arrived at indubitable truth, congealed as *dogmata* or *scientia*. In modern science the places of the *quaestiones* was taken over by hypotheses or theories, and the testing of these was taken over, not just by logic but by *empeiria* or experimental activity, coupled with mathematical logic.

THE RELATION OF PHILOSOPHY TO SCIENCE

At the M.I.T. Conference, technology was recognized as power, and the debate was about the ends to which that power has to be directed. But the nature of technology itself in relation to science was not very deeply examined there. The philosophical resources were sadly lacking at the conference and in the section that dealt with the question.

Here it seems important to look at Aristotle's notions of *causa* and *techne* which have, in modern times, undergone a substantial modification in the realm of science and technology.

Aristotle's notion of cause was a composite of four elements—the formal, the material, the efficient and the final. The causal element in a human artifact is the simplest example. Let us take an earthenware pot. The pot has its formal cause in the form of it that the potter conceives—a concept is always the formal cause, in this case the concept of a vessel—spherical, cylindrical or other, with an opening at the top and empty space within. The material cause is earth and water in the proper mixture. The efficient cause is the potter with his wheel—the wheel is not a material cause because it does not go into the pot as such; it belongs to the efficient cause, the potter with his/her skills and tools. The final cause is the end or use to which it is to be put—to contain liquids or solids. The human artifact is thus to be explained in terms of the totality of the four causes. Take away one of them and the pot would not exist.

In the case of a natural object the causal structure is most difficult to delineate. Take the clay itself which the potter uses. The material cause of earth or clay is simply *phusis* or nature. Its formal cause is (at that time of Greek thought) simply one of the elements composing nature—i.e., earth, air, water and sky. Today we would have to go into geological processes to explain clay. As for an efficient cause the ancients would perhaps speak of *phusis*, nature as an emergent reality actively bringing beings into being. And the final cause would at best be a matter of speculation.

The modern notion of causality has actually abandoned the idea of a formal or final cause for anything in nature. Science is an attempt to explain nature without reference to teleology or final cause. In fact there has been a firm dogmatic opposition to the introduction of purpose or end in the causal explanation of reality. Causality, devoid of final and formal causes, has been made to work in a mechanistic model of the universe up to a point. By developing the notion of mathematical laws of nature, we have dispensed with the Aristotelian structure of causality. The laws themselves are the efficient and formal causes. Until recently, final causes used to be entirely ruled out. But this meant the introduction of other factors besides the laws of nature—like the principle of life which contradicts some of the laws of physics.

Just to cite one example, the Second Law of Thermodynamics lays down the law that all matter is in the process of losing its energy content, moving from a more organized to a less organized equilibrium state (entropy). Life on the other hand shows the opposite trend— matter being organized to form more and more complex entities—with greater asymmetry or disequilibrium—the range is from virus to the human brain, in the bio-realm. Mechanism, then, has to add the principle of life or organism. We have tried to do this in science without introducing the notion of final causes.

We have tried to hold on to the mechanistic paradigm —with entropy (literally inner change) in matter and cosmos, and negative entropy in life as the principles, even to build up a theory of the evolution of the species. The inadequacies and deficiencies of this paradigm have been widely acknowledged. But in the absence of a better theory, we hold on to a neo-Darwinian or neo-Lamarckian theory of evolution.

Ilya Prigogine[12] and his associates have raised substantial questions about this mechanistic paradigm which disavows teleology. The Theory of Relativity has only established time as a fourth dimension of spatial reality. But the generic difference between time and space as dimensions is only now being highlighted. Prigogine, by establishing the existence of nonreversible processes in time, has demolished the notion of the uniformity of time. Unlike in space, in time one cannot move backwards and forwards. There is a point at which entropy stops and gives place to the negative entropy of life bringing order out of chaos. It is out of disorder and asymmetry that new forms emerge, bringing into being new asymmetric biological forms.

Morphogenesis is the name given to this process by Rupert Sheldrake. He argues that the mechanical paradigm based on causality as principle of explanation cannot account for the evolution of the species. Neither can the supplementary principles of vitalism or organicism account for the origin of new forms of being. As a biologist, Sheldrake[13] admits that laws of physics and chemistry, based in the mechanistic paradigm and causality conceived mechanically, can only go so far in explaining biological phenomena. Biology cannot at the moment abandon the mechanistic paradigm and explanation by causality without abandoning a great deal of what has been gained in biology. And yet, argues Sheldrake, the mechanistic paradigm faces huge hurdles in explaining mental processes. The computer model works—but only up to a point. "Breaking the genetic code is also a claim based on a mechanistic understanding of the genome. By analyzing the physical-chemical compositon of a chromosome, of genetic material (genome) or DNA or RNA molecules, we do not understand how the gene can contain a whole program for the growth and development of a member of a species. The physical-chemical difference between a human genome and that of a chimpanzee is, says Sheldrake, less that 1% of the total. He continues that the difference between the genome of Einstein and that of an average moron is even less. What is the mechanism by which a genome programmed for a particular organism operates? What is the process by which the programming itself occurs? And how does the programming make just that subtle change which brings about a new species?

Rupert Sheldrake argues for a programming that is built into the whole system of the cosmos—what he calls morphogenetic fields. But then every system that we know within the whole has small subsystems—animals, plants, cells, tissues, organs; even molecules and crystals and atoms are systems with subsystems. Each whole has its own programming, which is not derived from its subsystems. Even atomic particles are now explained in terms of subsystems called quarks. But in fact no system or subsystem can be explained just by its own constituent subsystems. The programming appears at increasing levels of complexity as we go up the scale of systems. And we do not know of the programming system in nonorganic systems. As we go up the scale, new forces and processes appear which cannot be explained in terms of the lower processes.

Sheldrake proposes as a biological theory the existence of morphogenetic fields, fields of unknown forces that generate new forms. The whole is a hierarchy of systems and subsystems which cannot be explained from bottom up. It can be understood only in terms of a directedness in the time-scale, and of forces operating which cannot be detected at the micro-scale, but are necessary for understanding the whole. And these forces are not a mere combination of the weak and strong forces we now know in dynamics or physics.

Biological evolution is a programmed process, not a random one. There may be indeterminacy in the programming of the whole itself. We know that the genome does not determine all the features of a developed human person. Some features are determined by environment, others by personal and social decisions within the human organism as it develops historically, as well as by historical, social and personal "accidents".

But Prigogine and Sheldrake together, one a physicist and the other a biologist, and the picture of the whole that emerges is curious. There seem to be formal and final causes operating in the whole, in the process of evolution itself. Prigogine tells us that, analogous to the speed barrier (the speed of light as the maximum possible), there exists an "entropy barrier". As the process moves from order to disorder, there comes a certain level of disequilibrium which then becomes the source of a new order, a new form. "Non-equilibrium is the source of order at all levels of existence".[14]

Can we say then that even in the evolution of human society there is a stage of nonequilibrium or chaos out of which a new order, a new form, emerges? Can we also say that this new order is not entirely preprogrammed, but that we actually participate by our historical decisions in determining the shape of this new form, at least in social formations, but also to an extent in the formation of our brains?

When we begin to say such things we have already moved from science strictly so-called into the realm of philosophy and faith. We have to understand as much as we can of the programming built into the whole. There science gives us the clues, but we have to step into philosophy and faith to complete our grasp of the nature of the programming and to decide on what our own role should be in the emergence of the new form of human social organism. Science is now incapable of telling us all that. It may do so tomorrow or the day after, let us concede in principle. But we have to make our decisions today, in the light of the imperfect scientific knowledge that we have. There philosophy and religious faith have to help us.

Which philosophy, which religious faith, one wants to ask. That is where the decision is needed today.

Sir Peter Medawar, the distinguished British winner of the Nobel Prize in Medicine (jointly with Sir Macfarlane Burnet) for the discovery of "acquired immunological tolerance which made organ transplants possible by overcoming immunological resistance in the human organism", has a perceptive essay in his *The Limits of Science*.[15] His argument can be summarized as follows:

> The Law of Conservation of Information which is a general law in logic and now in Informatics as a discipline, lays it down that 'no inductive generalization can contain more information than the sum of its known instances'.[16] 'The propositions and observation statements of science have empirical furniture only. In epistemological principle, they have all to do with ships, shoes and sealing-wax etc.' Since science is based on 'observation statements or descriptive laws having only empirical furniture, there is no process of reasoning by which we may derive theorems having to do with first and last things'.

If that is the limit of science, then science may be able only to tell us how things work. Its attempt to do so without positing a formal or a final cause in nature has itself come to grief as we clearly begin to see programs built into the whole at various levels, á la Prigogine and Sheldrake. One sees also more clearly at least one point where philosophy enters into science—the point at which the notion of causality conceived in mechanistic terms needs to be replaced by a richer notion of causality that includes formal and final causes. These causes lead us on into further questions which science itself cannot now investigate, because of limits of "empirical furniture which the scientific community has imposed on the methods of science.

The answer, however, is not to abandon science in order to espouse philosophy in a vacuum. Philosophy recognizes the limits of pure reason in a science based on repeatable experiments and precise measurement. The demand now is to shift our stance from that of science, and without losing perspective of the deliverances of science, to look at reality from a different standpoint, using methods other than those of modern science. In doing so, we need not abandon reason, but hidden or submerged capacities in the human person for reasoning of a different order have to be awakened. This awakening, called illumination by the religious, is itself a rightly philosophical task, but not within the limits set for reason's purity by "modern" philosophy beginning with Kant and Descartes.

Before we go further into this analysis of the task of philosophy in relation to science, we need to have a look at *techne* and technology from a philosophical perspective.

PHILOSOPHY AND TECHNOLOGY

"Technology in our modern sense is a new term coined, I am told, by Jacob Bigelow at Harvard in 1829. In the Greek language *technologia* meant something entirely different—the use of art and artifice in rhetoric, the cunning tricks of demagogical public speaking, the *techne* or cunning art of *logia* or discourse.

With Jacob Bigelow the word comes to mean the *logia* or discourse, about *techne* or the cunning art of manipulating things. The *techne* of *logia* now becomes the *logia* of *techne*. That may be a permissible departure, and the meaning of the word need not detain us here.

Heidegger's comment that modern technology is the last stand of an alienated Western metaphysics seems worth taking a look at. Let us view this transition through a few examples, some provided by Heidegger himself. Take a river for example and the way the human community views it. For a prescientific, premodern person, the river had a name and a personality. It was a friend, but one that you should not antagonize. If the river is angry, it may flood or dry up. It was fun to bathe in, it brought water for the crops, one could travel on it in rafts and boats, it provided fish for people; in short a good, beneficent and useful friend who formed the foundation of the life of the river-bank community and the valley civilization.

With modern science the emphasis changes. One studies the river in terms of its hydrodynamic and other qualities, with objectivity. It is no longer a friend to live with, but an object to be *studied*.

In technology, one goes one step further; one puts the river to *use*, mastering it with dams and hydroelectric plants, drains the energy from it and uses it for all kinds of other human purposes. From an object for study, it becomes a "resource for exploitation".

The same happens to the mountain, which makes the same transition from awesome friend to object of geological study, to resource for mining technology.

Heidegger sees in this process a chain of fundamental alienation. In the beginning nature was a friend, with whom and within whom one lived and functioned. We had a relation of responsiveness to beings and through them to the Being which manifested itself through beings. Human beings were dependent on nature. Nature could be terrifying in thunder and lightning, flood and earthquake; but nature was a dependable friend on whom the major activities of human activity like sowing, harvesting, hunting and fishing depended.

In the scientific objectivity of the early modern scientists there is an element of respect for nature. Nature's consistency was part of its dependability. And the main object of study in science was the structure of this consistency and dependability articulated through the mathematical laws of physics. But as modern science develops further, and as nature yields more and more of its secrets, there is an erosion of this respect for nature. Science becomes confident that it knows the beings and the laws that govern them. One need not look beyond the beings for Being which manifests itself through them. The beings are objects which can be studied and understood. The studying and understanding subject becomes thus dominant over the object.

But the object still remains object though an object that is already subject to the knowing subject. The mountain and the river are objects that we know. The mountain is "nothing but the product of geological processes", as is the river.

There is an element of gentlemanly supercilious contempt in that "nothing-but-ism" of the scientist—a looking down upon beings, though often with patronizing affection and wonder. But to think that the encounter with beings may be an occasion to heed the Call of Being does not appeal to the scientist. That would be subjective stuff—to be left to poets and artists. Scientists do not want to be contaminated with that kind of a subjective, feeling-based approach to reality.

The alienation of the scientist at this point is a double one—in relation to the object, and within the subject. The object is no longer a friend, but an object for dispassionate study. It is seen as the given, a law-governed "thing". As we make it reveal the mathematical laws governing its being, the subject becomes master of the object. Beings are no longer with us, in reciprocal relation to us as subjects; they are objects under our scrutiny. Within the subject the relational aspect of subjectivity is transformed into a kind of separation of the pure intellect from the sense of being-in-the-midst and relational dependence. "Pure reason becomes the dominant element in scientific subjectivity, and power over beings through pure understanding of beings in their mathematical law-structure becomes the central aspect of human subjectivity".

This alienation or cutting of the being-with relation of the human subject with beings then serves as the basis of the deeper alienation of technology, the desire to dominate, not just by understanding, but also by using the understanding for manipulation of beings in accordance with our interests and our will as subjects. The understanding also undergoes further alienation when the mountain is understood not merely as the result of geological processes, but as the source of timber or mineral ore. The timber merchant and the mining engineer develop a new understanding of the mountain as being-there (*da-sein*, in Heideggerian language) to be exploited for human purposes. The scientific knowledge about ore formation and tree growth now gives place to the technological knowledge of how best to exploit the tree and the ore.

The human subject now becomes no longer simply *under*-standing, but also *over*-powering, in relation to the natural object. The understanding becomes the platform for the overpowering. Science, through the understanding process, treats beings as object; technology, by the overpowering process, creates the machines and methods to use beings as resources for human consumption.

This is the metaphysics of the human "pure reason". It first subjects beings to itself by making them objects of the understanding through modern science. Then the human will proceeds to use the understanding of science to make beings merely resources for exploitation.

Modern science and modern technology are processes by which the human mind and will overpower beings and obliterate the Being that shines through beings, by making beings subject to human mind and will. God or Being becomes dead, and the human being seeks to take over as the being that knows and controls all beings.

In premodern science and technology also the human mind sought to understand and the human will tried to control. The simple building of a house or a wheel required such understanding and will. What has happened in modern science and technology is a change of stance of the understanding and willing self from a position *within* nature to *over* nature. Nature, which was previously inclusive of human beings, now becomes exclusive of the human.

The environmental crisis facing humanity has put the modern scientific-technological stance under question. The inclusive view of nature is commended by environmentalists, but modern science and technology are still groping in the dark for reestablishing humanity's stand among things and within nature as a friend, as one who is dependent on nature.

We have understood many things as a result of the environmental crisis. The first is that our very existence is relational. We do not first exist and then try to build a relationship with the environment. It is the evolution of our planet and the solar system that has made the biosphere or the life-world within the solar system possible. Humanity itself is a product of nature and the environment. The material world, in the process of evolution, creates life. In this sense nature is our mother, but we are still in the womb. We have not come out of it. We are within a structure of relationships with nature—the sun that warms and gives energy to our planet, the air we breathe, the leaves that create photosynthetic food from the sun for our bodies, the matter in our bodies taken from the earth, the light by which we see things. Take away any of these things and we can no longer exist.

We are thus not only produced by nature, but also continuous with it, conditioned by it, amidst it, still in its womb, integrally part of it now, though once it existed without us.

Modern science and technology have helped us attempt a nondependence on nature, making it subject to our understanding and will, thus dominating it, but not depending on it. Have we succeeded? If so, in what measure? Completely? or only partially? Whether complete or partial, have we had to pay a price for that emergence from the womb of nature and standing over against it? What should we now do?

These are the questions which modern science and technology raise, but for which they cannot provide the answer. These are important questions affecting our very existence. Especially so when our nonexistence, not only as the human species, but as the life phenomenon, is threatened with extinction by a nuclear holocaust.

The kind of philosophy which attempts to ask and answer these questions cnanot be merely an academician's game. Philosophy must first awaken itself and recognize the gravity and portent of its work for humanity.

The higher *techne* is not supplied by modern technology. The art or cunning or skill to find our way through the things which we understand and manipulate is not in the realm of technology but in that of true philosophy.

PHILOSOPHY AND THEOLOGY IN RELATION TO SCIENCE

Philosophy and Theology were not always as neatly separated as they are in modern times. The distinction was practically nonexistent in the non-Western traditions.

In the Hindu tradition for example there is no word for philosophy or theology, but one word for both—*darṣan* or vision. The distinction between faith or religion on the one hand, and philosophy on the other is a Western phenomenon, with its origins in scholasticism and particularly in that neat separation in Thomas Aquinas between philosophy as things we can know by our reason, and theology as discourse on what is revealed. Thomas himself has difficulty keeping this distinction neat, for in theology or the realm of the revealed there is interplay between what is objectively revealed and reason's analysis of that revelation. Even in Thomas reason is the unifying principle, reason reflecting on nature is philosophy, reason analyzing the revelation is theology.

In fact most languages do not have a word for religion as such—not even Latin or English, not to speak of Greek. *Religio* did not mean religion in our modern sense; it meant only "bound by a rule" and applied strictly only to the monastic as distinct from the mundane (worldly, not secular in the modern sense).

Hegel, in his esssay on "Believing and Knowing"[17] draws attention to the fact that the ancient oposition between Reason and Faith, between Philosophy and Positive Religion, has taken a new and altered sense inside modern thought. Philosophical thought was traditionally the task of throwing a bridge from the sensible world to that which lies behind and beyond that world, an act of human self-transcendence. But "religion" is also this throwing a bridge to the beyond. In fact modern science is also going beyond the surface of phenomena to the hidden "laws" behind them, thus also an act of self-transcendence.

It is also important to acknowledge the fact that the isms we have created in Western thought (e.g. Hinduism, Buddhism, Taoism, Shintoism, Judaism; however, at least in English, no "Christianism") do not correspond to any reality. These categories like religion, faith and theology or philosophy have arisen in the Western tradition, and are unjustly and unfairly imposed on other non-Western traditions. This idea that there is a category called "religions", understood in terms of belief systems into which all the non-Western traditions can be forced, is an aspect of Western cultural domination, resented by those in other traditions, but often reluctantly accepted because the Asians want to speak to Europe in the latter's own terms and categories.

The protest was courteously but clearly expressed by the distinguished Indian philosopher J. L. Mehta in his *India and the West: The Problem of Understanding:*[18]

> The two major traditions of Indian spirituality, Vedanta and Buddhism, are as much philosophical as religious traditions and are therefore neither purely philosophical systems nor religions, as these terms are understood in the West . . . So long as we insist on interpreting religious traditions such as these in terms of their believing-knowing dualism, they cannot be understood in their own essence.[19]

Professor Mehta criticizes the Western tradition for its missiological frame within which it seeks to deal with "other religions". The worst case, according to him, is Paul Hacker of Muenster, West Germany, who provides a clear elucidation of the Christian missiological approach—that of assessment, assimilation, appropriation and "utilization" of the wealth of their religions. Hacker defines "utilization (Greek *chresis*, or Latin *usus justus*):

> Utilization connotes, (1) that the assimilated elements are made subservient to an end different from the context from which they were taken; (2) that they can be taken over because some truth is contained or hidden in them; (3) that they

must be re-oriented in order that the truth might shine forth unimpeded.[20]

Mehta questions this desire for "utilization" as imperialist and acquisitive. He does not even approve wholly the method of Wilfred Cantwell Smith and Mircea Eliade, which seem free of missiological or Christian theological interests. These latter thinkers have a "planetary culture" framework, the vision of a global pluralism not tied to any particular theologies, but devoted to "an integral, universal way of thinking about *homo religiosus* and about human history" as in essence religious history. Eliade concentrates on myth, ritual and symbols, while Cantwell Smith focuses on religious history, but both in a global context. Mehta likes Smith more than Eliade. The latter seems to be pursuing a chimera, in its claim "to arrive, by the study of myth, ritual and symbols, at universal, objective, synchronic knowledge about human religiousness".[21]

Smith, on the other hand, seems to Mehta right in seeing religious traditions as growing wholes, constantly interacting and changing through that interaction. Mehta's approval of Smith stems from the latter's willingness to see the intellectual pursuit of truth in the world religions as an authentic aspect of all religions.

But there is need to be vigilant. *Religionswissenschaft*, even when purged of the Western cultural-imperialist *hubris* that characterizes such lofty thinkers like Hegel, Husserl and Heidegger, stands athwart the dangerous abyss of that scientific-technological alienation which we have discussed earlier. In objectifying religious traditions as a *fach* or discipline governed by the rules of science, we distort and reify them. With the technological temptation to "utilize" other people's religious traditions to enrich one's own, a further disastrous alienation sets in.

Mehta himself is fully aware of this danger, it seems. As a critical Heideggerian, he locates the danger in the framework (*Gestell*) of thought: "the peculiar constellation of man and Being that lies hidden, unthought, in technology as its characteristic mode of concealment". If Heidegger insists with some insolence on the totally Greek character of all philosophy and metaphysics so far, he would also go beyond to the planetary dimensions of human thought by unconcealing or unveiling the unthought element in Greek and Western thinking since Socrates— this question about the nature of thinking itself and its relation to Being and to beings in time.

Two basic ideas or insights seem to emerge from all this. First we do not mean by philosophy the kind of metaphysical speculation or ontological, linguistic or epistemological analysis that we have learned to call philosophy in recent centuries. This means we have to go back to the concepts of *Sophia* or wisdom as distinct from knowledge or science (*episteme, scientia*), in which the self and world are in communion with each other within the larger communion with Being in its integrity.

The second insight is that one cannot insist on a presupposition-less beginning for philosophy or a total separation between religion and philosophy in vision and reflection. The categories of science, philosophy and theology which we have developed in the Western tradition can no longer be held in watertight compartments.

This second insight becomes immediately problematic. It is with very great effort that we have come to carve out a realm of universal knowledge which is independent of geography and culture. The axiom is that science and technology are by nature universal; religion and culture are by nature local and particular. How can we then bring them together without destroying the integrity of both science and culture?

Clearly the answer is that science must retain its autonomy and freedom to explore without direct reference to divisive and unproductive philosophical and theological questions. That answer, however, seems a bit too simple. The argument here has been that science goes wrong when it becomes unanswerable to anybody. The cumulative impact of our argument so far is that science and technology are answerable to the human community. The community can question science/technology at several levels:

(a)   At the political economic level, the funding of

scientific and technological research and development has to be monitored, so that the weight of its use does not fall on destructive weapons, unnecessary consumption and unethical profit for the corporations; government or state as it is now constituted cannot have sovereignty in this regard; the public must come into the monitoring process and institutions have to be created to make this possible. This monitoring has philosophical implications, insofar as the ends to which science and technology are put cannot be regarded as scientific questions. They are moral ones which need philosophical reflection on what is good for humanity.

The situation thus calls for a further reflection on political economy, not only on how to create the institutions necessary for monitoring the financing of science and technology. The question will have to be put in reverse order—what kind of science and technology will foster a political economy that fosters justice and the equitable distribution of power in society.

Science and technology are the source of power in our society. That power is both unequally distributed and unjustly deployed. Bringing that power back under human control is just as crucial as the power of technology to control nature.

At this point the dialogue between physical and social scientists on the one hand and practitioners of political-economic power on the other seems to need philosophical mediation, so that thinking of the relation between the nature of the power of science/technology as forces of production and of the power of the political economy as the relations of production can proceed on a proper reflective basis.

(b) The scientific and technological community itself has a responsibility to examine the philosophical premises on which science and technology are based— the premises of scientific objectivity, technological utility and the domination of nature. We are doing violence to nature by our technology in that process violating our human integrity. The kind of subjectivity that emerges in the stance of treating nature as an object for study and as a resource for exploitation is that of a distorted humanity, out of creative and life-giving communion with nature and with the whole of Being.

Here the community of science and technology must enter into dialogue with the communities of art and literature, music and poetry, as well as of a quickened and renewed philosophy and religion. This dialogue should result in a reorientation for science and technology, but also in a revitalization in the other realms of art and literature, music and poetry, philosophy and religion. The variety in these latter realms would not thereby be eliminated. On the contrary, it is hoped that that variety would only be further enriched.

At this point the task of the community of philosophy and religion becomes even more arduous and demanding than that of science and technology. Philosophy and religion or technology have first to develop the tools and the competence to come to terms with what is involved in the human enterprise of science and technology. They have, in that process, also to reformulate their disciplines and reorient their concerns, so that they can provide, in dialogue with the practitioners of science and technology, a deeper perception of humanity's stake in science/technology.

Theology and philosophy have also the further responsibility to undertake the deeper reflection on the nature and function of these disciplines, and the role they have played in bringing humanity to the present stage. It is at this point that the plurality of philosophies and theologies can itself become a creative factor in entering into dialogue with each other, in the context of the issues raised by their reflection on what modern science and modern technology do to humanity. Philosophies and theologies now prevalent, with rare exceptions like Heidegger and Mehta, seem unequipped to take up this deeper analysis. Equipping them with the necessary competence for this task demands more than the dialogue among the various theologies and philosophies that are current. Dialogue should include perceptive scientists like Prigogine and Sheldrake.

Even when philosophers and theologians undertake dialogue with scientists, the dimension of political economy cannot be ignored or excluded. This makes the dialogue difficult to organize. Overcoming these difficulties should be one of the primary goals of special task forces within and without the academic community. The universities and study institutes of the world have a pioneering role to play here, possibly with the cooperation and support of international organizations like UNESCO and UNIDO, provided these latter are not allowed to suck the investigation into their cumbersome bureaucracies and thereby stifle them. Hence the importance of local and national initiative in a multipronged approach to the problem.

CHAPTER THIRTEEN

Science and the World of Living Religions

We have consciously taken a Christian perspective in our analysis of science, technology and philosophy. Some of our assertions may be acceptable to those who affirm other religious or secular traditions. But there is need for the other perspectives to enter into a critique of what is said here.

This need is not for Christians alone, nor just for Westerners. We can agree that the West has a special mission given to it by God—mainly to bring technology to full flower, to make a world of global communication and interaction possible and to sharpen theoretical understanding of this world and of human thinking and acting. Without that technology, the global transport and communication system and the theoretical grasp of human thinking and acting, we would not have been able today to enter into genuine global dialogue with each other.

The need for global dialogue is widely recognized today as a common need of all humanity. The feeling of urgency about it is still lacking, partly because the feasibility of such a dialogue was disputed even when it was practically possible. Besides, the Western interest put too much hope in liberal Capitalism or Marxist Socialism as the only possible alternatives. It was too lightly taken for granted that the great cultures of the non-Western world would be forced to capitulate to one of these alternatives. It was the very assumption of the invincible power of science/technology in a secular framework which put the dialogue with other religions and cultures rather low on the Western agenda.

What has happened? Far from receding and disappearing, the world religions are asserting themselves with a renewed vitality. They are prepared to enter into dialogue with the West, but in a context of rebuff and even hostility. They have come on to the ground cleared by Western culture. They have learned the languages and categories laid down by the West. The project for dialogue itself is seen as an attempt by the West to recognize the strength of the world religions on the one hand, and on the other to dominate these religions in this new arena called dialogue.

Dialogue is thus seen by the world religions of East Asia, as well as by the West Asian religions of Judaism and Islam, as a new battlefield with the West, where the war has to be fought with new weapons and new strategies. Perceptive thinkers of Asia like J. L. Mehta see this new "dialogue" and "hermeneutics" attempt of the West as a new try at overpowering and assimilating the rest of the world into its own greed-fired "being". It is a desire of the West to be "enriched" by assimilating "the spiritual universes that Africa, Oceania, Southeast Asia open to us", to quote Prof. Hacker, whom Mehta calls "the impeccable scholar of Advaita Vedanta".[1]

Sensitive thinkers in Asian religions are determined to resist this European or Western civilization in its new move to engulf and overpower them in the arena of dialogue. They know that they can fail in this battle if they adopt the thinking categories of the West and accept the Western standards of judgment which stem from the desire to understand in order to dominate, appropriate and manipulate. Historical factors have helped the West to develop this capacity to know, to possess and to work on things. If we approach the world religions with the same motive of knowing, possessing and "enriching" ourselves, then the West will decide what are the "values" in these other religions which are worth "appropriating".

The religions of the world are living entities, not dead cultures of the past. There are millions of people living and finding meaning in these religious cultural traditions. Of course a small percentage of these peoples have been assimilated into the Western culture, and now form the ruling elites of countries outside the West. They may even concur with the West in the modern Western cultural tradition, with its twin modes of liberalism and marxism, that they, the Western trained elite, can be arbiter about what is best in all religions. Dialogue of Westerners with Westernized Asians and Africans may even come to some easy consensus. That can be dangerous. To be foretold of this possibility is to be forewarned.

It is the Western civilization, which moved from the theological, through the metaphysical and the scientific to the modern technological, that has now established the Westerners as well as the Westernized non-Westerners at the heart of the homelessness and meaning-less-ness of a world civilization dominated by a scientific-technological elite. The Hegelian desire for *Aufhebung*, the principle of assimilation or swallowing and digesting, marks the onslaught of Western civilization in the world.

In the West there are thinkers who have shown us the untenability of the *Aufhebung* approach, not only to religions and cultures and ideas, but even to thinking about things—notably Heidegger, and following him, Gadamer.

If in our fundamental reflection on modern science and technology and what these are doing to us as human beings, Westerners or Christians invite the other religions in order to force them into Western thought structures, that reflection is foredoomed to failure. Raimundo Panikkar, who once wrote on *The Unknown Christ of Hinduism*, has now consciously abandoned that track, for it is a typically imperialist track, where we assimilate and appropriate Hinduism's best insights as our own by right, by claiming that they belong to "our" Christ. The West, and especially Christians, will have to start learning to say a new kind of "we" for whom there are no "they"—an inclusive we, embracing the whole human race, and willing to regard the Western tradition as one among our many traditions. Beyond Orient and Occident, beyond any narrow and exclusive we, all the peoples of the world have to learn to think of all human traditions as "ours", not giving privileged place to any—except occasionally to correct previous marginalization.

In such a dialogue where all the partners together say we, the "tribal" people have a special role. Precisely because they have refused to accept the Western desire to dominate and manipulate nature through modern science and the technology based on it, they have been marginalized to the point of near-extinction, and will need privileged attention. For these tribals, whom we despisingly call primitives, show us some aspects of that communion with the world to which we now aspire and of which we stand in desperate need. Their "primal vision" should now illuminate the horizon of the rest of us.

In setting up such a global dialogue of religions and cultures within which the question of science and technology has to be looked at again, the Christian and secular West may have a special role to play. This role is justified by the fact of their present privileged and central location in the flow of life—a privilege which has come to them by way of science and technology in part, but also in part from their aggressive-dominative colonial-imperialist rape of the world by which they generated that science and technology. The West may pioneer in developing that dialogue and in providing for its continuance. No one else at present has the power and the inclination to initiate or the wherewithal to maintain it. But the West's taking an enabling role should go hand in hand with their willingness to abjure a central or dominating role, for the rest of the world cannot at present trust the West.

AFTER HEIDEGGER WESTERN THOUGHT GOES ITS OWN WAY

Heidegger is right in his claim that philosophy in the Western tradition of questioning Being by standing apart from it is Greek and exclusively Western, yes, the very basis of Western civilization, and the matrix of modern science and technology. If the dialogue, however, is to take place within that civilization, for the interest of that civilization and within the categories of that civilization, the rest of the world has reason to be cautious, maybe even scared for fear of being again swallowed up.

Heidegger has also indicated the way out of this doomed path of theology, philosophy, science. The point that he makes is seldom grasped. To grasp it, the Western thinker has to lay down the attitudes developed in the course of Western development:

> The present day planetary-interstellar world situation is, in its essential origin that can never be lost sight of, through and through European-Western-Greek . . . What alters it is capable of doing so only out of that which remains saved, of its great origin. Accordingly the present day world-situation can receive an essential alteration, or at least a preparation for it, only from that origin which has become fatefully determinative of our age.[2]

The fate of Heidegger's thinking, when Heidegger himself was assimilated into the ongoing structure of Western thought, warns us of the peril that awaits us if we do not open our eyes and ears before we launch the great global effort for a global concourse of religions. Heidegger has given rise to the new "hermeneutic epistemology" which seeks to replace the scientific method or Aristotle's *episteme theoretike*. Gadamer is the central thinker in this new development of Western thought. He can be credited with the honor of bringing into light a new hermeneutic on the basis of Heidegger's analysis. Gadamer refused to reduce language to a mere game or sets of games, each with its own rules. That was Wittgenstein's failed effort. Gadamer gave validity even to the scientific method, which was reeling under the discovery that scientific hypotheses were subjective creations which in a sense preceded empirical knowledge, projections of foreknowledge which the scientist then tested out experimentally.

But the difference in tone and purpose between Heidegger and Gadamer is both noteworthy and important for our looking at the world dominated by modern science and technology as well as for our approach to the global community of religions. Heidegger is a sage, a prophet who uses the analysis of being to open people's eyes and ears to the Call of Being. Gadamer, by using that same Heideggerian analysis, creates a new *method* of understanding. This was what Heidegger resolutely refused to do. Heidegger's purpose was to disclose the total structure of Greek thought, as it stands outside of being and questions being, in order that people may see the fundamental *nonvalidity* of the enterprise in which Western civilization has been involved for twenty or more centuries. Gadamer, on the other hand, validates that enterprise, by going beyond Kant in *describing* subjectivity in terms of tradition, horizon and *wirkungsgeschichte* or the history of effects on the subject. Kant's subject analyzed as pure reason, practical reason and aesthetic judgment was ahistorical. Gadamer delineates the historical element in human subjectivity and thetic intentionality (a la Husserl).

Heidegger, as sage and prophet, knows well that he stands within the Western enterprise of the subject seeking to position oneself outside the world and overpowering the world by knowledge and technology; yet he seeks to give us a novel perspective on that Western enterprise, shows us how it obscures being, and asks us to wait in the clearing, wait to heed the Call of Being. Gadamer, on the other hand, simply gives us a more adequate and satisfactory way of continuing the Western enterprise. Heidegger exposes the sensitive reader to tear off the mask from the Western enterprise and see the harsh visage of nonbeing lurking behind the mask. Gadamer simply does some plastic surgery to that harsh visage. The result is that the silence necessary for heeding the Call of Being becomes drowned out in the new hermeneutics. Gadamer exemplifies the point of Heidegger's warning against using *da-sein* analysis to produce a new metaphysics which stays within the Greek-European *Ge-Stell* or structure. As J. L. Mehta rightly points out:

> Heidegger's thinking has little to do with "cultural synthesis" or with the notion of a "planetary culture", or with the idea of a "universal philosophy" for the man of today, gathering together the complementary insights of the philosophies of the West and the East. His thinking is post-philosophical, in the sense of being no longer "metaphysical" and no longer operating on the presuppositions implicitly at work in all "philosophy".[3]

This is an important warning for the global dialogue of religions that seems necessary as a prelude to the re-orientation of science and technology. In order to heed that warning adequately, we have to listen to criticisms of Gadamer advanced by Juergen Habermas, which can just as well be applied to Heidegger, nevertheless.

Habermas accepts Gadamer's critique of the Kantian enterprise which took the adult modern (German?) mind as the standard for universal mind, and went on to analyze the categories of understanding of the pure reason on the basis of that assumption. Gadamer showed us the infinite variety that can exist in the consciousness of different individuals. Each mind has its horizon of perception drawn by the history of the effects playing upon the individual mind. One's education and training, one's knowledge of facts, one's taste and temperament, one's cultural conditioning, the tradition within which one stands and many other factors determine how far and how well the individual subject can see, i.e., the extent of the subject's present horizon. This horizon of course keeps changing with further experiences and further training.

It is this analysis of human subjectivity in Gadamer that Habermas criticizes as too individualistic. Habermas extends Gadamer's project. Starting once again with Kant, Habermas asks more questions about the formation of the individual mind. And without going to Heidegger, who makes the same mistake as Kant in analyzing the human *dasein* in terms of the Western experience as understood by the adult mind, Habermas resorts to Hegel's analysis of self-formation or *Mensch wurdung*. Epistemology, even a hermeneutic epistemology, says Habermas, takes too many shortcuts in this analysis of *Mensch wurdung* which we wrongly translate as humanization.

Consciousness as subject has to be analyzed as a process, not to be too quickly and simplistically understood as the result or cumulative history of happenings to which the consciousness has been subjected. Ego-formation has itself a dialectic, as Hegel taught us, says Habermas.

Acknowledging Gadamer's contribution to the dynamization of Kant's subjective ego which was a static given with an unalterable category structure, Habermas accuses Gadamer of insufficient analysis of the subjective consciousness. Hegel in the *Phenomenology of the Spirit* told us quite clearly that the object-in-itself, the *Ansichseiend* can be experienced by the subjective consciousness only as for-us-being, as *Fuer-uns-Seiend*, as being for consciousness. This Gadamer has recognized as inevitable.

But in analyzing the history of the development of the subjective ego, we should start with Hegel, says Habermas in *Knowledge and Human Interests*.[4] The human consciousness grows dialectically, i.e., by progressive negation of previous consciousness, induced by new knowledge. The role of negation in consciousness is not totally ignored by Gadamer, but not analyzed deeply enough. The negation-affirmation or anti-thesis-thesis dialectic is not to be seen as the result of external affects on consciousness, but the very nature of truth, as Hegel saw it, reflected in nature and consciousness alike. This self-formative process of the human consciousness has at least three dimensions—(a) the socialization process of the person as affected by culture, education, training, etc., (b) the universal history of humanity which has resulted in the particular culture in which the person was socialized, and (c) the experience of the particular person through the basic forms of reflection, namely science (*Wissenschaften*), religion and art.

From the commonsense evolutionary point of view, later adopted by Marx, mind is something that evolves out of nature. For Hegel, what makes nature evolve to produce mind is itself Mind, the Absolute Idea which operates through the laws of the dialectic in nature. Mind is the *arche* (*Erstes*) of nature. Mind is the Idea existing for itself and in itself, and which strives to overcome the gap or alienation between subjective and objective, in nature as well as in the human consciousness.

Marx reasserted the evolutionary or scientific principle that nature is the ground of the human mind as subject and of the world as object, as well as of the interaction between both subject and subject and subject and object.

The subject-object relation is not mainly one of knowledge, but of labor, handling the external world and transforming it for human ends. This labor however is social labor, of organized societies cooperating with each other in the act of molding the objective world for human ends. Animals too interact with the environment, but they usually *adapt* themselves to their environment. Man is different, in his tool-making capacity as *homo faber*, his knowledge and his language merely assisting him/her in tool-making and engaging in social labor along with others to transform the world. The genesis of man is not in language or understanding, but in tool-making and transforming the world.

Meschwerdung or human becoming and consciousness formation as an aspect of it are simply the genetic act by which man enters his own history, which is a new stage of natural history. The mechanism of this human becoming is socially organized labor. People produce in order to reproduce their own life as a species interacting with nature though part of it.

In this Marxian analysis, Habermas detects a defect and proceeds to rectify it. Marx identified two strands in the process of human becoming—namely the forces of production and the relations of production. The first is covered by science and technology and the second by political economy. Human subjectivity and its goal-choosing activity are subsumed under the relations of production. Marx failed to pursue the analysis of human subjectivity and the role of human subjects in the process of development. Scientific-technological as well as political economic research is an essential aspect of the objective complex of natural history we seek to analyze. But the methods applicable in science-technology and political economy are insufficient for analyzing the very process of human subjectivity, and its role in *directing* the process of natural development.

The process of human becoming, through interaction with nature and through interrelation with other human beings in social labor, has a third level, that of critical assessment of human social activity at both levels, i.e., in science/technology and in political economy. The history and particular development of the human subjectivity which engages in critical assessment and seeks understanding should itself be subjected to critical analysis. Such critical analysis, however, is not a mere theoretical activity oriented merely to *understanding*. The theoretical activity has itself an interest that goes beyond mere understanding. Critical theory is, or should be, oriented to critical praxis. Description of human subjectivity should lead to its practical implications.

Habermas criticizes Gadamer for making reflection or understanding an end in itself, and thereby for overvaluing the place of reflection and understanding. Reflection has an emancipatory interest, the hope for a new enlightenment that leads to a new emancipation and a new reorientation of emancipatory practice. Reflection may yield social norms for human subjectivity. But only practice can work out the specifics of the application of these norms under varying conditions of social development.

Habermas seeks a "reconstruction of historical materialism"[5] by taking apart the elements in Marxist theory and putting them together again. He comes out with a three-level structure in place of Marx's two-level analysis of Forces of Production and Relations of Production. The additional level is *Ideologiekritik*, or criticism of ideology. His three levels are in fact a reversion to and refurbishing of the three levels of Kant—pure reason, practical reason and critique of judgment. His contribution is in specifying the three different forms of validation or justification appropriate to the three levels. Truth claims in science are validated by testing in external nature, providing grounds in experiment. At the level of political economy or rightness claims, validation has to be provided in norms of interactive moral values, justification in terms of values like justice, freedom and equality.

At the third level of value choices or *Ideologiekritik*, justification, according to Habermas, relates to the theoretical questions raised by praxis at the other two levels. The process requires diagnosis or reflective analysis of the historical process by which we have arrived at the present, and a prognosis in terms of real possibilities desired to be achieved in the future. One studies the present state of society, not for contemplative satisfaction, but to see its full possibility and to help advance the praxis leading to the realization of this possibility. In studying society, we have a moving object and we can understand it only in terms of where it came from, what forces now direct it, and what alternate trajectories are possible for it. From the "is" and the "was" of society we should move towards the "ought" of society, and towards the praxis necessary for the ought to be realized.

But how are these "ought" choices to be made? At this third level, Habermas rejects the phenomenological method of Max Scheler and Nikolai Hartmann, Max Weber's ahistorical sphere of norms, and Karl Jaspers' "forces of faith". Neither would he accept Sartre's existential decision or any idea of arbitrary choice of values. None of these are valid substitutes for reflective reason using the hermeneutic method.

Habermas would argue for two fundamental principles which undergird his "meta-critique". The first principle is that of the primacy of practical reason over theoretical reason. We want to know in order that we may act. "Action is the pre-supposition of knowledge, wanting-to-act is the presupposition of being-able-to-know", as Garbis Kortian puts it.[6] The second principle, derived from the first, is the unity of the practical reason and the theoretical reason, of theory and praxis, of theory *for* praxis. Habermas' quest thus becomes radical. In place of a universal theory of understanding, he speaks of a "universal pragmatics", the science of a theory for praxis. This is basically Marx's view that the human being is constituted in the process of *Handling* or *Stoffwechsel*, molding material reality to his interests, the theoretical being an aspect of this handling of nature. The subject-object interplay takes place through theory and practice, which together then shape the human subject.

But Marx's error, according to Habermas, was in regarding social production as the total realm of theory and praxis. In the realm of social interaction for social production, institutions of domination and exploitation arise, developing a particular kind of reflection which morally justifies that domination and exploitation. This is what Marx called legitimation or ideology, e.g. bourgeois ideology. Now Marx's critique which leads to the unmasking of this ideology does not fall strictly within the categories of social interaction necessary for social production, but is at a level above both. Marx's reflection was not just phenomenological description of a social process, but a critical unmasking by a sort of social psychoanalysis and exposure of the interest lurking behind particular ideologies. Marx tried, however, to assimilate his own critique to the theory of social interaction for the sake of social labor.

Habermas wants to see another interest that guides knowledge. Without denying knowledge's interest in action, Habermas suggests another interest—namely that of emancipation from the constraints of untruth. This was what Marx was trying to do himself, but he did not reflect on the nature of the interest which guided his own critical reflection. Following the line laid out by his colleagues at the Frankfurt School (Adorno, Horkheimer, Marcuse), Habermas carries the distinction between technical reason whose interest is production and ontological reason whose interest is emancipation from untruth, into the distinct discipline of *Ideologiekritik*, defined as Universal Pragmatics, or a universal interpretation of the processes of formation—ego-formation, family formation, socialization, formation of institutions like the state and the political party. This is metapsychology, metaphilosophy, meta-critique.

This discipline evolves only through communication without constraint, whose main tool, language, has itself to be psychoanalyzed to bring out its tricks of concealing the truth. This capacity to communicate without constraint which he calls communicative competence is what is constitutive of the human species, rather than tool-making as Marx taught, or understanding as Gadamer proposed. Communication without constraint, however, has its final goal, truth as consensus through psycho-analytically and critically exposing all pseudoconsensus.

> The idea of true consensus demands that the participants in a discourse be capable of distinguishing between being and illusion, essence and appearance, is and ought, in order to be able to judge competently the truth of propositions, the veracity of utterances and the rightness of actions. In none of these dimensions, however, are we able to name a criterion which would allow an independent judgment on the compe-tence of possible judges, that is, independently of a consensus achieved in a discourse. The judgment of competence to judge must in turn appeal to a consensus, for the evaluation of which criteria were supposed to be found. Only an onto-logical theory of truth could break out of this circle. None of these theories, however, has as yet survived examination.[7]

HEIDEGGER, HABERMAS AND THE DIALOGUE OF RELIGIONS

This laborious and highly inadequate discussion of Habermas had two purposes: first, to exemplify the change in tone and purpose between Heidegger and those who consciously took some cues from him, like Gadamer and Habermas. These latter try to take the Western tradition beyond the crisis to which Heidegger pointed. But in going beyond, they have forgotten his appeal to heed the Call of Being in silence.

Secondly, the discussion on Gadamer and Habermas has something very important to say to us about the pitfalls in a wrong approach to the global dialogue of religions. If religions are now to be analyzed purely in terms of the Gadamerian or Habermasian category structure, they will again be sucked into an enterprise in which the Western categories dominate and distort the genius of non-Western cultures.

It is interesting that neither Heidegger nor Gadamer have paid sufficient attention to some of the huge distortions and monstrous cruelties inflicted by Western culture—not only on non-Western peoples, but also among their own neighbors, the Jews. Why does the human *dasein* which Heidegger so brilliantly analyzed, make no reference to the Jewish pogroms and the gas chambers in the concentration camps? Why did not Heidegger recognize, as part of his analysis of human *dasein*, the fact that it was European civilization which gave rise to two world wars? Why do neither Heidegger nor Gadamer make any reference to the criminal acts of colonialism and imperialism, which are part of the *wirkungsgeschichte* of the late 20th century European mind? Why is there no reference to the mad armaments race and the abominable phenomenon called nuclear stockpiles? Why is there no reference, even in Heidegger's appeal to heed the Call of Being, to the cry of the poor and the oppressed in the world today? How can we ask for the silence needed to heed the Call of Being, when we have not yet heeded to the cry of the oppressed?

The Dialogue of Religions that wants to look at science/technology and the problems it raises for humanity will have to look at the whole in a different perspective than that of the affluent West which sees "homelessness" as the central reality of its world civilization. The other kind of "homelessness", namely that of not having a proper roof over one's head for many millions, is just as much a feature of the civilization—however Greek-European it may be—which the West has created.

In other words the Global Concourse of Religions, in its effort to seek a different orientation for science/technology, will have to see it from the perspective of the poor and the oppressed. Unfortunately, however, the people who will take part in an interreligious dialogue today have very little direct experience of the world of the poor and the oppressed. Even the people who represent the non-Western religions come from the affluent elite within their own religions. They seldom have the direct awareness or experience of the literal home-less-ness and food-less-ness of their own people, when they try to deal on a global basis with the spiritual homelessness of the world civilization made in the West.

"HOMELESSNESS" AND RELIGION

Good religion is always one that cares for the whole. Not only for the whole of one's nation, not just the whole of humanity, but for the whole of the created order of beings. Good religion sees humanity as the "Shepherd of Being". It therefore cares for the poor and the affluent. Christ cared for the poor and declared that "theirs is the kingdom of God", but he was compassionate for the rich too. In the story of the rich young Elder of the Jews, who told him that he had kept all the laws of the Torah from his youth, Mark tells us that Jesus "looked upon him, loved him, and told him—'You lack one thing, go, sell what you have, and give to the poor. . .and come, follow me'".

Religion has to learn to look with compassion on the plight of the affluent, their actual imprisonment to their wealth created through the domination and exploitation of science and technology, and their resultant "spiritual" homelessness, their inability to enter the kingdom of God through their wealth and their science and technology.

"Go, sell what you have, and give to the poor" has today to be understood in a new socioeconomic and international sense. This is not today a matter simply of the affluent selling what they have and giving it to the poor. It requires a sacrificial effort on the part of the affluent to enter into a New International Economic Order in which wealth and power are not concentrated in the hands of a few. This means accepting to live with others in an order of economic relations (trade policy, banking, finance, debts, investment, control of the means of production and all that) where no one seeks privilege and each one cares for all while all care for each.

In such a world alone we can seek to abolish the spiritual homelessness of the affluent and the material homelessness of the poor. This latter is partly created by the conditions which led to the former. Modern science is characterized by the principle of domination and modern technology by the ethos of exploitation. Even human beings are in the process reduced to "human resources", for "development". Investment in education itself is conceived as an economic move for promoting "development", for providing the scientific and technological personnel necessary for economic growth.

Development there has been, but largely of a middle class, who have become a vested interest, having power over the process of development through their control of production, government and media, and therefore in a position to make development tilted in the direction of their own class interests.

In this lopsided world, the religons play a negative as well as a positive role. The negative role of religion is poetically expressed by the Indian historian-journalist Romesh Thapar in an eloquent passage in the special issue of the Indian Government's Planning Commission Journal *Yojana*:

> Paste on your Hinduistic caste-marks and your symbolic *tikkas*, you sinners of modern India—yes, even within the imitative five-star culture you have spawned! Deck yourselves in saffron turbans of vengeance, against the teachings of the Gurus even though you lack the courage to speak the truth about what really goes on! Trim your beards, Maulvi-like, to demarcate yourselves even as you chant parrot-like your prayers in languages that are unfamiliar and pass on this "culture" to those who see the "light" in following you! Become the followers of a nonviolent Buddhist dharma, for it does not really prevent you from taking a tooth for a tooth or two for one![8]

The negative role of religion has been pinpointed under four headings:

(1)   its misuse by politicians to further their own ends and enhance their power;

(2)   its resistance to reason, progress and enlightenment and its promotion of superstition and ignorance;

(3)   its causing violent conflicts between communities divided on the basis of religion; and

(4)   its hypocrisy in putting on airs of piety and while mouthing high principles like love, freedom, tolerance, etc., practicing hatred, discrimination and exploitation and sanctifying these by the forms of piety.

Clearly the global concourse of religions should have as one of its main purposes the formulation of a program to combat these negative aspects of religion—all of which the founders of all religions would themselves repudiate. It is thus not an attack on religion to pinpoint these negative aspects and fight against them. It can spring out of a commitment to religion itself. It does not require the four-fold program proposed by the editor and contributors of this special Independence Day issue of *Yojana* with its title *WHY LIVE WITH THIS NON-SENSE!*

(i)   arrest religion and see to it that it remains confined to homes and places of worship alone;

(ii)   divest government of its traditional role in selling religions and make it work to promote rationality and humanism;

(iii)   base all education on rational explanation of human growth and teach rationality and humanism in our educational institutions; and

(iv)   use government media, *not* as it is used, but use it to promote humanism and rationality.

The positive aspects of religion, however, are too often overlooked in the pursuit of reason, progress and enlightenment.

One of the points overlooked is the source of the values. Let us for the moment concede the point on which Marxism and Habermas agree: that values are derived from human experience and not brought from heaven.

On what aspects of human experience yield values, we have differing approaches in modern reflection. One of the most intriguing lines is provided by sociobiology as advocated by E. O. Wilson[9] and Richard Dawkins[10] Starting from a superb analysis of Ethology or the science of animal behavior, and after being corrected in the course of the debate, Wilson comes to the moderate view that genetic traits developed in the process of (animal) evolution *condition* but do not necessarily *determine* human behavior. This need not be questioned. Wilson, however, has failed to account for the fact, pointed out by Marshall Sahlins in 1976, that the element of culture was the mark of a break in continuity between animal and human behavior. That is the point underlying Marx and Habermas. Marx says that human history makes a beginning with the possibility of the human being's capacity to make tools and thereby make nature into culture; this tool-making capacity and interaction through it with inherited "nature" constitutes human nature. Habermas would contend that the tool-making capacity is inseparable from the evolution of language and intersubjective communication.

Dawkins, on the other hand, tried to put all behavior on the gene that seeks to survive through the species, and undergoes mutations in the process of that survival struggle. Wilson, himself, particularly in debate with Dobzhansky, has accepted that genetic determinism accounts only for part of human behavior, the other part being determined in the process of *cultural* (as opposed to *biological*) evolution of the human species. He may disagree with Dobzhansky on the relative roles of biology and culture, the latter arguing for more influence by culture, while Wilson holds on to more influence by biology.

Wilson repeated in 1975[11] his warning against the "naturalistic fallacy of ethics, which uncritically concludes that what is, should be". He only insists that "the 'what is' in human nature is to large extent the heritage of a Pleistocene hunter-gatherer existence". A genetic bias, when discovered, cannot be used to justify following it. "Genetic biases can be trespassed, passions averted or redirected, and ethics altered."

The question needs to be asked then about the source of origins of these genetic-bias-altering norms in cultural development, and the role of religion in it. The good modern humanist would argue, with August Comte, that religion was integral to human cultural evolution. He would argue that when we have arrived in the age of science, we can abandon religion and metaphysics, to develop a "scientific humanism". This argument for a "scientific humanism" is quite different from arguing for "humanistic science" such as Abe Maslow and his school would plead for. We need to have a look at the claims of scientific humanism.

If the project for a "scientific humanism" is viable, then we need no global dialogue of religions. We should, on the other hand, like the editor of India's *Yojana* magazine, ask government to promote scientific humanism, banning all discussion of religion from the public sphere. In a sense Habermas' Universal Pragmatics is such a plea for a rational humanism which would be completely independent of religion. Habermas has no difficulty acknowledging that values like truth, openness and free communication may have existed once in a religious milieu; but he argues that these values are now self-evident.

Habermas' anthropology has these two foundations of a modified marxism: *labor* and *language* as constitutive of the human species. He would then go on to find two elements in the rationality that guides labor and language— namely instrumental or *technical reason* and communication competence. He then goes on to delineate three spheres in the activity of reason through labor and language—namely those of the natural sciences, of the human sciences and of judgment norms.

He is not arguing that science can be the basis of humanism. The universal pragmatics of Habermasian humanism calls for three different sets of norms or validation principles for the three levels. The norms of the physical sciences cannot be applied at the level of the human sciences or at the level of value judgments.

The question that Hegel asked Kant can be asked of Habermas as well. Kant was questioned about the method of categorical analysis being used for delineating the category structure itself. Pure Reason cannot by its own method define the categories of Pure Reason.

Similarly, one can ask, by what principle is Habermas going to validate the validity norms themselves? He can only resort to a consensual basis of social acceptance of the validity norms. That consensus we certainly do not have yet.

HUMANISM AND THE NEED TO GO BEYOND

Corliss Lamont in his impressive work, *The Philosophy of Humanism*[12], has attempted a synthesis of humanistic philosophy. He criticizes George Santayana, the American philosopher, who wrote *The Life of Reason*,[13] for later weakening "his philosophy by adopting an esoteric doctrine of essences, which much like the old Platonic ideas, are supposed to subsist in an eternal realm apart from the regular course of nature".

But even Lamont has to affirm some basic values: e.g. democracy, and free speech.[14] He goes on to analyze ten areas in which democracy has to be maintained: politics, civil liberties, racial and ethnic relations, economics, organization, social relations, culture and education, religion and philosophy, sex relations, international relations. He does not say by what scientific principles he arrives at these categories of a scientific humanism.

He has an appendix, the *Humanist Manifesto*, signed by leading western thinkers like John Dewey and Edwin A. Burtt. But the Manifesto turns out to be a plea for a "religious humanism", a religion which affirms that the universe is self-existing, that "intelligent enquiry" is the way to determine values, that the end of man's life is the complete realization of human personality "in a free and universal society in which people voluntarily and intelligently co-operate for the common good".

It is an abdication of intelligence, it seems to the present writer, not to ask the question about where such values are derived from. Hector Hawton, the editor of *The Humanist* in Britain, seems also unaware of the fact that "the premise that man must solve his problems with his own unaided resources and that "Humanists have to work out for themselves their answers to the quesitons which have inspired the great religions"[15] are both dated and obsolete. These heroic humanists are unable to recognize the force of Gadamer's argument that the prejudice against prejudice is a prejudice inherited from the European Enlightenment. The Western Humanist-Christian debate remains at a low philosophical level, unfortunately on both sides. Hector Hawton understands that "Modern Humanism derives in the main from the French Revolution and the Enlightenment". But he fails to see how the absolute role given to Reason in the French Revolution and Enlightenment is a burden that reason has proved itself unable to bear. Modern science which powered the Enlightenment has begun to recognize its limitations, but Humanists are slow in catching up with these new insights. British humanists loudly affirm "morality" over against "religion", but are unable to see the problems of reason in revealing the sources of morality. Habermas is unique in his affirmation that the only possible validation of moral judgments is consensus through unhindered communication. We shall not try to analyze further the understanding of truth implied in that affirmation.

At this stage all we can do is to point out that in the end there is no indubitable norm that is given to us. We have to accept responsibility for the decisions we make, and no norm is going to absolve us from the consequences of our actions and our thought. Humanism's growing strength in the last 200 years of European history is no validation of its truth. Even Marxist humanism is unable to get to the consensus that can hold it together. Marx's discovery that under capitalist conditions of production, philosophy itself has been reduced to an ideology, i.e., false consciousness, remains valid to this day. The question whether also under a powerful socialist state with state-owned means of production and the awareness of being besieged by class enemies, Marxist philosophy can turn into a defensive and obscuring ideology, as a false consciousness which hides truth, has now been raised within Marxist circles. The "fetishism of commodities" which Marx located as the source of alienation functions powerfully even in Marxist societies today and has not been eradicated. Humanists and Marxists, as well as religious leaders, are prisoners still to this fetishism, and to the love of power and domination. Liberation still remains a project, even in Marxist societies.

Ferdinand Toennies, the German sociologist, has illuminated the distinction between society and community, between *Gesellschaft* and *Gemeinschaft*. The first is a structure, legally enforced; the second is a being together in freedom and love. Even Habermas remains caught in this duality. At the level of science-technology and political economy he argues for *Gesellschaft* or predetermined structure. At the level of value choices he has recourse to community, free communication and consensus.

Communism also argues for a struggle through *Gesellschaft* to *Gemeinschaft*, a stage beyond the class struggle where structures give place to free spontaneity.

Our Global Concourse of Religions is a proposal to begin some aspects of that future *Gemeinschaft* or spontaneous community, even now, in the midst of our struggles and our disagreements. The precious insights and achievements of secular or religious humanism have a contribution to make in this Concourse. If the religious thinkers get together simply to analyze their own agreements and disagreements, they will be performing a useful academic exercise, but little beyond that.

The Global Concourse of Religions must include the Humanist and the Atheist and provide not only for their total freedom of expression, but also to heed their legiti-

mate criticism of religion. Religions themselves need renewal and refinement today, and these not just for the
survival of religions, but for the emancipation of humanity
and to promote the emergence of a renewed humanity.

COMMITMENT TO A NEW HUMANITY

Commitment to a New Humanity has to be the guiding force of the Global Concourse of Religions and Ideologies we are proposing. Without that commitment, our
very attempt to reorient science-technology and philosophy will prove self-serving and worthy of contempt. There
are many conferences and assemblies of world religions
being held all over the world, but these are marred by negative attitudes towards certain sections of humanity.
There the religions seek to justify themselves in self-
glorifying propaganda that lacks this concern and commitment to the new humanity.

The Global Concourse of religions and ideologies has
to be a school of learning for all who participate, and for
all human beings. It is at this point that the problem of
elitism rears its ugly head. If it is the future of humanity
that is at stake, then that future cannot be trusted to a
con-course of intellectuals. It must take up not only the
concern *for* the masses, but also provide for the concerns
of the masses to be expressed and dealt with.

This is difficult today to provide for, unless educational institutions and systems, the mass media, and the
political parties and organizations, the trade unions and
peasant organizations also play an active part. We will
need a concerted effort that includes serious intellectual
reflection, but also uses art forms and literature, music and
poetry, demonstrations and festivals with mass participation and the creative contribution of the masses in forms
most congenial to them. Science-technology would have
to be dealt with dialectially. Its positive achievements,
especially where it beneficially affects the masses, will have
to be highlighted. A negative attitude towards science-
technology itself will have to be scrupulously avoided.

At the same time the captivity and misuse of science-technology itself should be revealed in forms which the masses can grasp.

The religions themselves would have to highlight their nobler aspects. At the same time the captivity and misuse of religions will have to be uncompromisingly manifested by the practitioners of each religion, careful not to offend and lead to communal clashes and religious conflicts.

Where such exposures turn out to be caricatures arising out of scurrilous motives, there should be freedom to respond and reply in peaceful ways. This applies equally to the exposures of science-technology as to the religions. The overarching concern should be, one never tires of repeating, the care for the present and the future of our human race.

What will such a concourse achieve? This is hard to foretell. It will have its impact, if effectively set up, on all realms of experience, on all institutions and societies. It will awaken humanity, let us hope, to a new awareness of its own plight. It should kindle new hope, new striving for a global community. It should dispel despair and hatred and promote care, love, understanding, and self-criticism as well as a new humility so essential for true community. It should dispel the pall of gloom that lies over large sectors of the masses and spur them to hopeful and creative participation in the emergence of a new global community.

It will also have to be watchful about global and national vested interests who in the guise of promoting and financing it will seek to take it over for their own self-regarding and self-serving ends of advertisements and market manipulation. The necessary safeguards will have to be built into the structure of the Concourse itself.

The Concourse will not be an alternative to the political, economic and social struggles for emancipation that will go on side by side. Insofar as these struggles form a necessary part of humanity's emancipation from cultural, economic and political domination and exploitation, the Concourse itself will help the masses to understand the true nature of these struggles as part of its own work and not as a hindrance to it.

The concourse will definitely meet with opposition, not all of it from antihuman forces. The Concourse will be patient in listening to such criticism and eager to change its ways when the criticism is valid.

The Concourse will strive to be parapolitical, but will not shun politics as such. The political-economic realm, where the most important decisions affecting humanity are made, will remain a central concern of the Concourse, but its basic approach will be that of care and compassion for all.

What specific role will religions themselves play in this Concourse? Would it not be better to keep the Concourse free from all religions, so that religious strife will not mark or mar its work? Why not choose rather the Secular humanist frame?

The answer is clear. Multitudes in the world still live where the religious consciousness is the strongest motivating force. The point is to emphasize the nobler and ennobling aspects of religion and to emancipate the religions themselves from their negative stranglehold on the people. This applies to all religions.

What does one mean by the nobler aspects of religion? I can only illustrate at this point from two religions I know best—Hindu Dharma and Christian Commitment. One exemplifies the East Asian tradition. One has to be very brief at this point. Values are more often caught than taught. And so long as values are assimilated, the source from which they are derived matters less.

THREE POSITIVE CONTRIBUTIONS FROM THE HERITAGE OF THE HINDUS

Hinduism is a vast corpus, with voluminous writing and innumerable perspectives. Compared to it, the tradition of Western thought up to the 15th century looks very limited and manageable. The present writer claims no comprehensive knowledge of the classical Indian tradition, and comes to it with humility and awareness of incompetence. Three strands in this rich and kaleidoscopic tradition are here merely mentioned, as illustrations to the generation of values and perspectives by religion—values which we can of course secularize and integrate into a modern humanism. If we need to trace the richness of these concepts, however, recourse has to be made to the living tradition itself as well as to the writings it has bequeathed to us.

When I look at the Vedic tradition, the earliest strand in the written tradition, I am fascinated by the concepts of *rta* and *yajna*. The universe is an ordered whole with its own rhythm and organismic unity, like a chick hatching out of an egg, growing, dynamic, but needing care. That is the basic meaning of *rta* which is not so much a concept, as a vision, a *darsana*, a perspective that guides action. But in this cosmic order in which the whole is related to each part and each to the whole, certain beings endowed with freedom, the divine beings (*dēvās*), the demonic beings (*asurās*) and human beings (*narās*), have the responsibility to care for this ordered whole and the ability to disrupt its order and harmony by their self-willed actions. The actions of free beings can care, tend and promote the order and harmony or by not caring, upset it, bringing disorder and disharmony. When this is done, the order has to be restored by sacrifice or *yajna* in which gods and humans cooperate in a ritual act. The genius of sacrifice is self-immolation to the whole. The focus is not on the salvation of the sacrificer, but on the tending and caring for the whole. For the sake of the whole, one offers oneself in sacrifice. *Yajna* thus restores the disrupted harmony.

This vision and perspective is easy to distort, not only in thinking, but in the practice of the priestly class who later became the professional sacrificers who knew the secrets of the *mantras* or the sacrificial formulas to be recited to make the sacrifice effective. But in the basic vision, the outlook and attitude is noble—self-giving and self-sacrifice for the sake of the whole, not a preoccupation with personal salvation.

In modern science too this vision persists, though in a faint form. The paradigm of cosmic wholeness and dynamic interrelatedness slowly comes back to the vision in modern physics. What is done in one part, says modern physics, affects the whole and other parts far away spatially.

In physics again, we are coming to see that the scientist as observer does not stand outside the cosmos, but merely enters into a relation of dialogue, of theory and experiment, in order to learn how nature works in relation to us and how we can work in relation to it. Fritjof Capra in his *Tao of Physics* gives us a poetic vision of the correlation between the order in the universe as revealed by physics, and the *sakti*-driven cosmic dance as perceived by early Taoism and Hinduism.

Even more important is the concept of sacrifice as a human value, and the absence of preoccupation with personal salvation in Vedic religion. *Yajnō bhuvanasya nābhi*, "sacrifice itself is the navel of the world", says the Rg Vēda and goes on to say that sacrifice is the true character of the human being.

These are not concepts. When expressed in words they sound pale and unconvincing. Ritual was the deepest experience of early humanity, and there the most profound perceptions are experienced. Can we not pick up these noble visions and generate new rituals in which to embody them? If entrusted to reason, it will become, to use an Indian expression, "a flower garland in a monkey's hands". Reason seems to have lost its capacity to wonder, to adore, to surrender in worship.

Take a second strand of the Hindu tradition—one that has been calumnied by supercilious Westerners like Albert Schweitzer to be world-negating. The *Gita* is the jewel of the Indian tradition and *Bhakti* its beauty mark. No talk about personal salvation here, no call to blind belief, no simplistic reduction of religion. In a tradition whose richness cannot be expressed in words, still the most vital of the Indian traditions, the great sage Sri Aurobindo attempted the following summary of *Gita* teaching:

Reposing one's mind and understanding, heart and will in Him (the Brahman) with self-knowledge, with God-knowledge, with world-knowledge, with a perfect equanimity, a

> perfect devotion, an absolute self-giving, one has to do works as an offering to the Master of all self-emergence and all sacrifice. Identified in will, conscious with that consciousness, that (*Tat*) shall decide and initiate the action.[16]

It is difficult to condense that statement, for it does not come from analytic reason, but is rather poetry coming from deep and rich experience of reality. To translate this into the pedestrian categories of Humanism would be sacrilege. It is a poetic invitation to an experience.

That experience lies out of the horizon of the modern secular humanist, who in his unhumanist skepticism may decry that experience as "mystical" or "superstitious", both of which probably mean much the same thing to him. The value of that experience is not a topic for discursive discussion among those who have never experienced it. All the values that can heal our societies can come out of such an experience and the perspective underlying it. Let technical reason stand back and, if necessary, walk away. But it has no right to ride roughshod into this domain with its relentless dissecting and dominating passion. The Concourse of Religions should make it possible for the reverent to reflect on something that does not so readily fit into their awareness of human experience.

I will take as a third example the very Vedanta or Advaita tradition so little understood and so miserably caricatured by the discursive reason. The Hindu Vedantin speaks out of an experience. His truth is not one that is arrived at by discursive reason or analysis, but by hearing, with heeding, meditation and discipline (*sravana*, *manana* and *nidhidhyṣa*). It is knowing indeed, but not the dualist discursive reasoning to which the West has become accustomed in the recent past. It does not proceed through the stages of objectification, analysis and synthesis, nor through the more modern technique of hermeneutic exegesis.

Under the expert guidance of one who has already experienced it, a *guru*, the disciple goes through a long period of self-discipline in withdrawing the tentacles of the mind constantly reaching out to things, in keeping the sharpness of one's mind without using it as an instrument to objectify and dominate the outside world. The disciple enters through nonreflective meditation into levels of the mind that the ordinary humanist or religious person does not even suspect to exist.

At long last there comes illumination where the duality of subject and object gives place to the *experience* of a nondual unity of knowledge at a higher or deeper level. In that experience the difference between knower, known and knowledge yields to a nondifferentiated knowledge which unites the three into one; there the disciple experiences the fact that the self is identical with the higher self of Brahman, and that same higher self is what is manifested to our senses at the lower level as the world of multiplicity. True philosophy in this tradition consists in the pursuit of this experience. The theoretical reflection that follows upon this experience of praxis is secondary. It is the experience, the realization, the illumination, the *siddhi* (attainment) of *mukti* or liberation-emancipation, that is primary. Taking the theoretical reflection apart by discursive reasoning does not lead to truth. For those who have not practiced and attained this emancipation, this project remains weird and illogical. But to the *Siddha*, the one who has attained, the certainty of the experience needs no validation by criticism or consensus. It is self-certifying and leads to *ānanda*, joy or beatitude of being in the truth. The truth is not one that the knower *possesses*, but the knower *becomes* the truth, unshakeable and unwavering.

No humanist critique of the discursive Western epistemological-critical-hermeneutic enterprise can do justice to this experience. It lies beyond and beneath that enterprise. The theoretical analysis of Universal Pragmatics has not taken account of this experience, because the experience has not been part of its praxis.

The experience itself leads to a new perspective on reality, a new attitude towards human beings, animals and all beings as manifold manifestations of the One who is beyond all Being and is yet identical with one's own self— the higher self. Care, compassion and respect for others should be the natural outcome of this experience. The *Siddha* then devotes the rest of his life drawing others to the same experience. The theoretical reflection or philosophy of Vedanta is not a justification or validation of the knowledge attained, but merely a witness to it, an invitation to it.

We have very briefly and all too inadequately sketched three major strains in the Hindu tradition, merely to show how religious experience goes beyond mere discursive or analytical-synthetic reason. One could do the same with other illustrations from Buddhism, Jainism or the Tao. We could also do the same with the West Asian traditions of Judaism, Christianity and Islam. We proceed to sketch in three strains in Christianity, again not aiming at any completeness of analysis and description, but merely to point to a dimension, to witness to another level, of the religious experience itself, which cannot be reached by the various varieties of humanism.

THREE STRANDS IN THE CHRISTIAN TRADITION OF ASIA

What follows is in the nature of a personal testimony from within the tradition to which the present writer is committed, without closing oneself to other traditions both religious and secular with which one wants to be in constant dialogue. I write as an Eastern Orthodox Christian, born and nurtured in it, but fairly well and for long exposed to other Christian and non-Christian traditions, and grateful for that exposure.

The Primacy of the Eucharist

For an Eastern Christian, the Eucharist, commonly called Mass, Lord's Supper or Communion Service in the West, is central to the Christian life. The Bible as the Word of God is heeded primarily in the context of eucharistic worship.

The Eucharist is not a "sacrament" or a "means of grace" as the West understands it. Sacrament is a Latin word with its own history and tradition. It meant first the pledge of a promise or the seal set to a contract. We Eastern Christians can understand it in that sense, as the enactment in our presence with our participation, of the New Covenant (sometimes wrongly called New Testament) between God and humanity in Jesus Christ who is inseparably united God and human person.

But our basic understanding, which can not be exhaustively explained in words, is that of a community act on behalf of all humanity. The community of the Church, as united with Christ the only Mediator, lifts up the sacrifice of the created order, in a thank-offering to God. The faith-community united with Christ stands as the priest of Creation before God, and makes its offering of thanksgiving on its behalf to God. The elements of the offering, bread and wine, are products of God's creative activity but transformed by the human activity of milling wheat and baking it, of gathering grapes and taking the juice out them. These divine-human products of bread and wine are the symbolic offering of the created order. The creation offers itself to God in thanksgiving, both for creating it and then redeeming it from evil and death.

It is this self-offering of the Created Order to God, through and with Christ's offering of himself to God on the Cross, that constitutes the first movement in the Eucharist. This first movement itself is a response to God's loving activity in creation and redemption, an activity which is proclaimed again and again in the reading of the scriptures and in preaching. The Church, as the body of Christ in this world, does it on behalf of all creation as its thankful, joyful response in repentance to God's activity of creating and redeeming. In fact it is Christ himself who does it through the Church, for He alone is the true High Priest.

In this first movement of the Eucharist, the world of science and technology, of political economy and philosophic reflection, is lifted up and offered to God as the fruit of our labors and of the bodies and minds that God has given us for this labor. This is where the Christian ultimately locates science-technology, political economy and value choices, in creation's self-offering to the Creator.

Science-technology, political economy or ideology choice will not rule the Eucharistic offering. It subsumes them and lifts them up to God as offering.

The second movement in the Eucharist is God's self-offering to us in the Body and the Blood of Christ. By God's giving Himself to us he feeds us, nourishes us, empowers us to go on working in the created order, ennobling and sanctifying it as God's manifestation of himself in love, wisdom and power. This renewed community then goes forth to its various labors whether it be in production, ruling or reflection. Science/technology and political economy as well as reflection fall within these labors done by Christians and others in God's created order. Others may not recognize the activities of science or politics as a manifestation of God's creative activity operating through human beings. The Christian community understands all human activity, in fact all the dynamism and vitality in the created order, as God's own creativity operating in freedom also through human beings who have the freedom not to know or to deny the source of their own creative activity.

The Eucharist is the central and normative experience through which the community lives, and authentic theoretical reflection on it can be based on authentic practice. There is no way the humanist with his limited categories does justice to the experience or to reflection on that praxis.

The Centrality of the Human

The humanist speaks of the centrality of humanity in a sense different from that of the Christian. Corliss Lamont quotes with approval Edward P. Cheyney's definition of humanism as "a philosophy of which man is the centre and sanction". Lamont supplies his own definition: "a philosophy of joyous service for the greater good of all humanity in this natural world and advocating the methods of reason, science, and democracy."[17]

Here the individual human being stands as his/her own authority, accepting only reason, science and democracy as qualifying that authority. The humanist serves and advocates. The humanist has himself/herself as authority, as agent and as end—with a few qualifications like the "greater good of all humanity in this natural world" and advocacy "of the methods of reason, science and democracy".

To the Christian this appears a highly inadequate conceptualization, to which the Habermasian project of a universal pragmatics with three levels of validation seems infinitely superior in acceptability. In Habermas there is a movement from individual to community, and a better awareness of the limits of reason and science. In the matter of democracy both visions strike this particular Christian as rather simplistic and not sufficiently differentiated.

The Christian sees the centrality of the human species in quite a different way, and cannot give in to the Humanist perception. Humanity is central in its relational existence—with God as the ground of humanity's existence and with the world as the given structure of its body-soul existence. It is precisely the denial of the first pole of human existence that makes Humanism as a philosophy unacceptable to Christians, especially Eastern Christians.

In the thought of Gregory of Nyssa (ca. 334-395), the foremost philosophical thinker of the classical Christian Church (largely and till recently very much ignored in the Western Church), humanity is not just in the middle or center, but is the *mediator* par excellence between God and his creation. The emergence of the first human beings in the chain of evolution, which Christians call the creation of humanity, was the terminal event of biological evolution. Humanity is the full flowering of the created order, as it comes into being and moves towards its fulfillment. The universe reveals its full nature in bringing forth the human community. The cosmos becomes self-aware in humanity, but this humanity subsists only as a fruit on the tree of the cosmos, on the tree of life.

It is precisely this mediating role that makes human mediatorial existence problematic.

The word used in Eastern patristic thinking for this mediation is *methorios* or *methoria*. The word actually means "frontier", the zone between two countries which participated in both countries, so to speak, in those days of no clearly marked boundary lines. Twilight is a frontier between day and night. Even the new Gadamerian hermeneutic concept of "horizon" is a *methorios*—a frontier between what we can know and what we cannot know by virtue of our historical limitations. For the Christian fathers Heaven itself is just a frontier—a border zone between what is visible or accessible to our senses, and what is not.

Aristotle spoke of *methorios* as the frontier between degrees of being—for example between organic or animate and inorganic or inanimate.[18] Gregory of Nyssa, on the other hand, does not use the term for that distinction, and but once for humanity's frontier existence between the sensible world and the intelligible world. Gregory's main use of the term *methorios*, however, connects it with human liberty; humanity is at the frontier between good and evil. For Gregory it is a misunderstanding to construe this frontier existence as between matter and spirit, as Philo of Alexandria did.[19] The decisive frontier is that between good and evil, which is also the frontier between being and nonbeing, between life and death.

In Christ, however, a new frontier has appeared—that between the Creator and the Creation. Christ, as we Christians understand him, partakes of both. He is both Creator (as God) and creation (as a human person).[20] Cyril of Alexandria, living two generations after Gregory of Nyssa, and to whom one must attribute the final clarification of the nature of Christ in the classical Christian tradition, put it thus:

> Christ is the *methorios* between divinity and humanity, being the coming together (*sunodos*) of the two in one.[21]

Cyril goes on to say that we human beings are now conjoined to God (*sunaptometha*) in Christ as the *methorios* between the divine and the human. It is in this new humanity inaugurated by Christ that modern science and technology have been born; but humanity still being at the *methorios* between good and evil, our science/technology, our political economy and our philosophical reflection are all capable of good and evil.

Here the Humanist optimism about science, reason and democracy has to be tempered by the Christian realism which insists that all our institutions are capable of good and evil, and humanity as *methorios* between good and evil has to be vigilant and watchful, and should strive to bring these institutions into the domain of the good.

The centrality of the human, says the Christian over against the Humanist, cannot be taken for granted. It is a double centrality—that between good and evil, and that between Creator and Creation. The Humanist temptation is to discard the second centrality in which humanity is responsible not only to itself.

This temptation to repudiate God and to take over the Creation as one's own is the monumental and primordial temptation. The environmental crisis itself is a result of succumbing to this temptation. But equally disastrous is the temptation of the religious leadership to take over the place of God and to tell other people what to do. It is humanity, and not just the Church or the religious leadership, that has to be the presence of God in the world, the shepherd of creation, caring and tending it through science/technology, political economy and value choices. The essence of human freedom is in accepting this contingency of humanity's existence and calling, and in creating, as God's presence, that which reflects the glory of God which is the glory of humanity also.

The Community of the Spirit

The Christian individual does not pose himself as the final authority, choosing what he/she wills. The Christian recognizes his/her triple belonging—(i) to the Creator to whom he/she owes his/her being, (ii) to the Creation in which he/she is set and on which he/she depends for his/her sustenance and (iii) the Community of the spirit living in Eucharistic communion with God and with other human beings. The third belonging, which non-Christians do not understand or always approve, is the arena in which the first and the second belongings are experienced in Eucharistic communion. For Christians, especially for the Eastern Tradition, to belong to this community is not simply to be a member of a club or a social organization. It is a community that spans space and time, for it is part of their vivid experience that they belong to an organism whose foundations were laid in the death and resurrection of Jesus Christ, and which has been growing through the generations; death does not take the departed out of this community. They too are living members of this community of the spirit, though not present visibly in the present.

But what constitutes the essence of this community is not the presence of the departed. Its being is ground in Him who is the *methorios* incorporating Creator and creation; it is also the community in which God is present in a special way as the Holy Spirit. The function of that presence is to prompt the community to do its task—to be the mediating presence between the Creator and the creation, offering up the life of creation to the Creator in thanksgiving, and promoting the presence of God in the world through the forms of the good expressed in love, wisdom and power, in value choices, political economy and science/technology. This spirit pours forth love, wisdom and power, beauty and truth in the world, as well as in the Church. To discern the work of the Spirit in the world and to promote it is the calling of the Church. The Church is differentiated from the rest of creation only by its conscious recognition of its foundation in Jesus Christ, by its discernment of the Spirit in the world, and by its continuous entry into the presence of God as the priest of creation, carrying offerings and intercessions. The Church constantly invites others to participate in this community by repentance and faith, by baptism and the Christic anointing. Those who heed are so received. But those who do not heed remain just as much in their caring and compassion. They, that is those who remain outside this community of the Spirit, have every right to call the community to task, when it ignores the way of the Spirit and uses its freedom to follow the way of evil.

To conclude

The foregoing was not offered with any dogmatic arrogance, but as part of an invitation to join in the global concourse of religions and ideologies. It is the present author's deep conviction that the redemption and renewal of science/technology, political economy and philosophical reflection need not and, may I say it, *cannot* take place without the participation of the religious communities. He knows that the religious communities are inadequately prepared for such dialogue, because they have often ceased to care and so failed to equip themselves. He also knows that participation in such a global concourse is eminently dangerous for the religious communities, especially in view of their inadequate preparation. It is quite possible that we may lose some of our best minds and spirits in the course of the dialogue.

This writer has no illusions about the comparative merits of present-day religious thought and secular thought. The definitely superior quality of secular thought, both liberal Western and marxist Western, may lead some people to abandon their religious loyalties. My plea is that the religious communities should take this risk, if they care more for the created order than for the survival of their own communities.

I would also plead that the secular thinkers show some patience with us. We promise to learn, but give us a chance by exposing your ways of seeing and thinking to us in such a global concourse. Let us together be committed to the care for the created order, so imperiled by the presence of evil in science-technology, in political economy and in value-reflection. Let us commit ourselves to banish militarism and the arms race, the nuclear stockpiles and our huge national armies. Let us commit ourselves not to use science-technology or political economy for oppression, domination and exploitation. Let us commit ourselves on behalf of humanity to turn the course of its development from evil to good, from destruction to reconstruction, from ugliness to beauty, from falsehood to truth, and from bondage to freedom, from gloom to hope, from boredom to joy. Let us do it together.

NOTES

CHAPTER ONE

1.   The phrase science/technology is used to denote the complex system in which pure science and applied science interpenetrate and operate as a single phenomenon in present society.

CHAPTER TWO

1.   *Oil and Gas Journal* (December 25, 1978); *Science* (April 14, 1978).

2. Solar energy can be used in three major ways:
   (a) space-heating.
   (b) water-heating.
   (c) electricity-generation.
   A fairly big home of Western standards with 30 sq. meters of mirrors or collectors can manage 30 to 70% of its space-heater needs depending on location in nontropical zones. The market for space-heating is now $50–$100 million. By 2000 A.D. it is to be 5 billion.
   Electricity is more difficult. Thermal conversion by focusing sunlight on water to produce steam to run a turbine needs 100 sq. meters of mirrors to produce about 10 kw. This is very expensive. A 300 meter high tower with 5 sq. km of mirrors will be necessary for small towns. Electricity is needed to pump water that high. The system cannot work when the sun is down . . . Photovoltaic cells now cost $15.00 a watt at noon on a sunny day. Synchronous orbit satellites are more difficult. The technology is yet to be perfected. People expect something by 2025 A.D. If more money is put into research a more economical technology may emerge by 1980.

3.   There are those who think that there is no great problem here. For example, Professor David Rose of MIT, who is a learned and thoughtful advocate of the peaceful use of nuclear energy, told us:

> The technology for disposal of nuclear wastes is in relatively good shape. For example, the Swedish proposal to encase them in lead and titanium jackets (and copper, if the spent fuel is to be entombed directly without any reprocessing) and then to emplace them in geologically stable granite formations with bentonite packing looks good. Eventually, disposal in the seabed may be even better; the North Pacific Plate appears to be exceptionally stable and geographically predictable.

The problem however is that people do not always trust the experts. They suspect that the experts are also human beings, and despite extreme care and responsibility, are susceptible to make the mistake of leaving out or not being aware of certain factors. There is some basis for this suspicion in the record of past performance.

CHAPTER THREE

1.   For further study of the issues involved in bio-ethics, the following books are useful: Beauchamp and Walter, L*Contemporary Issues in Bio-Ethics* (Encino, CA: Dickenson Press, 1978); Joseph F. Fletcher, *The Ethics of Genetic Control* (New York: Anchor, 1974); Paul Ramsey, *Ethics at the Edges of Life* (New Haven: Yale University Press, 1978); Ramsey, *Fabricated Man, The Ethics of Genetic Control* (New Haven: Yale University Press, 1970); Warren T. Reich, ed., *Encyclopaedia of Bio-Ethics*, 4 Vols. (New York: Macmillan, 1978); T. A. Shannon, ed., *Bio-ethics* (New York: Paulist Press, 1976).

2.   Preston N. Williams, ed., *Ethical Issues in Biology and Medicine* (Cambridge, MA, 1969), p. 12.

3.   In "On Controlling the Scientific and Technological Revolution" in *Dialectics and Humanism* (Spring 1979,Vol. VI, No. 2): 90.

CHAPTER FIVE

1.   The increase in population also led to pressure on land, hence to higher rents, hence to higher wages.

CHAPTER SIX

1.   At the College of Notre Dame du Lac in 1965, the present Notre Dame University celebrated a "centennial of science" in commemoration of that event. See F. J. Crosson, ed., Science and Contemporary Society (University of Notre Dame Press, 1967).

2.   For example, Prof. Hanbury Brown, in his address at the MIT Conference on The Nature of Science, cited the conventional description of science: "Science, viewed as a process, is a social activity in which we seek to discover and understand the natural world, not as we would prefer or imagine it to be, but *as it really is.*"

3.   For an interesting treatment of the various images of science in the West, please refer to Carl Hamburg, "Science and Institutional Change" in K. H. Silvert, ed., *The Social Reality of Scientific Myth* (New York: American Universities Field Staff, Inc., 1969). The present writer is indebted to Dr. Hamburg's article.

4.   One of the earliest works of Maritain was entitled *Anti-modern* (Paris, 1922). In the preface to his more mature work on *Science and Wisdom* (London, 1940), he takes the view that "science is in itself good and noble" (p. 6) but only "in the sense in which one perfection is inferior to another perfection."

CHAPTER SEVEN

1.   Many professional beggars in India have been found to have bank accounts or significant sums of money otherwise stored away.

CHAPTER EIGHT

1.   J. Maritain, *Integral Humanism, Temporal and Spiritual Problems of a New Christendom*, trans. Joseph Evans (New York: Charles Scribner's Sons, 1968). It was orginally published under the title *True Humanism*.

2.   Op. cit., p. iv.

3.   Ibid., pp. 6-7.

4.   Soren Kierkegaard, *Concluding Unscientific Postscript*. Eng. text in Bretall (ed.), *A Kierkegaard Anthology* (Princeton, 1951), p. 214.

5.   *Sacramentum Verbi*—An Encyclopaedia of Biblical Theology, J. B. Bauer, ed. 3 Vols. Vol. 3 (Herder and Herder), p. 869 ff.

6.   *Fifth Theological Oration:* XXVI (Eng. Tr. The Nicene and Post-Nicene Fathers, Second Series, Vol. VII), p. 326.

7.   Ibid., XXIX. *Patrologia Graeca, Vol. XXXVI: Cols. 166-68 (author's translation), NPNF, p. 327b.*
     The neuter pronoun is used since the Spirit (to pneuma) is neuter in Greek; the personal element has to be read into the pronoun which is neither masculine nor feminine.

8.   See Paul Verghese, *The Joy of Freedom* (New York and London, 1967) and *The Freedom of Man* (Philadelphia, 1972); Paulos Gregorios, *The Human Presence* (Geneva: W.C.C., 1979).

9.   One needs to make it clear that one does not speak about Creation, Sin, Fall, etc., in clear, rational, scientific language. A mythic language is indispensable for the discussion of such concepts. The reader should seek to see what symbolic sense he or she can get out of these myths, but not treat them analytically in a rational sense.

CHAPTER NINE

1.   The expression "heaven" has been interpreted by the ancient fathers as that which lies beyond the horizon of our senses; not as the top floor of a three-storied universe, or as space above the vault of the sky.

2.   The formulation of Hume's question is from Karl Popper's *Objective Knowledge—An Evolutionary Approach* (Oxford, 1972) which was a response to Thomas Kuhn's *The Structure of Scientific Revolutions*, 3rd Ed. (Chicago, 1970).

3.   Bertrand Russell, *History of Western Philosophy* (London, 1946), pp. 698 ff.

4.   *Objective Knowledge*, p. 81. See also Popper's *Conjectures and Refutations—The Growth of Scientific Knowledge* (New York: Harper. Original edition N.Y., London, 1962.)

5.   *Objective Knowledge*, p. 263. His contention is that knowledge does not begin from nothing, but essentially that knowledge grows by modification of existing knowledge or mental predisposition.

6.   A. Tarski, "Der Wahrheitsbegriff in den formalisierten Sprachen" in *Studia Philosophica* (Vol. I, 1935): 261 ff. Eng. tr., "The Concept of Truth in Formalized Languages" in A. Tarski, *Logic Semantics, Metamathematics* (1956), Paper VIII, pp. 152 to 278.

7.   *Against Method, Outline of an Anarchistic Theory of Knowledge* (London: New Left Books, 1975—339 pp.).

8.   Lakatos, Imre and Musgrave, Alan (eds.), *Criticism and the Growth of Knowledge* (Cambridge: Cambridge University Press, 1970, reprinted 1976).

9.   Op. cit., p. 92.

10.   Popper provides some criteria for measuring the closeness to truth or verisimilitude of any given scientific theory in terms of its "truth-content" and "falsity-content", "content" here being meant to stand for all the consequence statements entailed by that theory.

11.   See Elie Halevy, *The Growth of Philosophic Radicalism*, Eng. tr. Mary Morris (Boston: Beacon Press, 1966), p. 433.

12.   It can be argued that what is characteristic of so-called modernity in philosophy is precisely this divorce between science and metaphysics, which then leads to the acceleration of secularization.

13.   Dilthey's work was preceded by that of the philosopher-historian Droysen who tried to make a neat distinction between Explanation (*Erklarung*) and Understanding (*Verstehen*), the first being the task of the natural sciences and the second that of history.

14.   Hans-Georg Gadamer, *Wahrheit und Methode* (Tübingen: J. C. B. Mohr, 1960); Eng. tr., *Truth and Method* (London: Sheed and Ward, 1975), pp. 239-240.

15.   Gadamer says that three things are necessary to examine a prejudice critically: (a) recognize it for what it is and that it exists; (b) objectify the prejudice by stating its nature, so that we can look at it; and (c) find a good prejudice which will help us understand the prejudice we are examining.

16.   For Habermas' critique of Gadamer's *Truth and Method*, see Jürgen Habermas, et al. (eds.), *Theorie—Diskussion, Hermeneutik und ideologie-kritik* (Suhrkamp Verlag, 1971), pp. 45-56, as well as *Philosophische Rundschau* (Tübingen: Beiheft 5, 1967).

17.   G. A. Kursanov, "The Problem of Truth in the Philosophy of Marxism" in *Philosophy in the USSR, Problems of Dialectical Materialism* (Moscow: Progress Publishers, 1977), p. 203. Academician Kursanov defines truth as "the process of the reflection in human consciousness of the inexhaustible essence of the infinite material world and the regularities of its development, which at the same time implies the process of man's creation of a scientific picture of the world emerging as the concrete historical result of cognition that is constantly developing on the basis of socio-historical practice which is its highest criterion" (ibid., p. 205).

18.   Kursanov, op. cit., p. 226.

19.   V. A. Lektorsky, "The Dialectic of Subject and Object, and Some Problems of Methodology of Science," in *Philosophy in the USSR*, op. cit., p. 109.

20.   "Philosophy investigates the same world that is investigated by the specialized sciences. But it cognizes more general connections and relationships than the specialized sciences, which study certain particular spheres of phenomena . . . Every science investigates a qualitatively definite system of laws in the world—mechanical, physical, chemical, biological, economic etc. There is no specialized science, however, that studies laws common for the phenomenon of nature, the development of society and for human thought. It is these universal laws that form the subject-matter of philosophical cognition . . . Academy of Sciences, USSR, *The Fundamentals of Marxist Leninist Philosophy* (Moscow: Progress Publishers, 1974), pp. 32-33.

21.   T. I. Oizerman, "The Problems of the Scientific Philosophical World Outlook" in *Philosophy in the USSR*, p. 38. The Western view of science's alleged independence of ideology, according to the Marxists, serves only to obscure the integral relationship of Western science-technology to the capitalist ideology, and thus to keep Science/Technology a servant of the market economy system.

22.   Karl Marx and Frederick Engels, *Collected Works*, Vol. III (Moscow: Progress Publishers, 1975), pp. 332-333.

23.   Ibid.

6.   Garbis Kortian, *Metacritique, The Philosophical Argument of Juergen Habermas*, Cambridge University Press, 1980, p. 76.

7.   Habermas, "Verbereitende Bemerkungen zu einer Theorie der Kommunikativen Kompetenz in *Theorie der Gesellschaft oder Sozialtechnologie*, Frankfurt: 1971, p. 135. Eng. Tr. Kortian, *op. cit.*, p. 125.

8.   *Yojana*, Aug. 15, 1985, p. 7.

9.   *Socio-biology: The New Synthesis*, Cambridge, Mass.: Bellknap, 1975 and *On Human Nature*, Cambridge, Mass.: Harvard University Press, 1978. See also analysis of the socio-biology debate in Gunther S. Stent (ed.), *Morality as a Biological Phenomenon: The Presuppositions of Sociobiological Research*, Los Angeles and London: University of California Press, revised ed. 1980, and Arthur L. Caplan (ed.), *The Sociobiology Debate*, Harper and Row, 1978. See also Michael Ruse, *Sociobiology, Sense or Nonsense*, Boston: Dordrecht, 1979 and Charles Lunsden and E. O. Wilson, *Genes, Mind and Culture*, Cambridge, Mass.: Harvard University Press, 1981.

10.   Richard Dawkins, *The Selfish Gene*, Oxford University Press, 1976.

11.   *New York Times Magazine*, Oct. 12, 1975. See *The Sociobiology Debate*, p. 267.

12.   Fifth Edition, Revised and Enlarged, New York: Frederick Ungar Publishing Co., 1965.

13.   New York: Scribners, 1954.

14.   Lamont, *op. cit.*, p. 47.

15.   *The Humanist Revolution*, London: Barrie & Rockliff, 1963.

16.   Sri Aurobindo, *Essays on the Gita*, Pondicherry, India: 1972, p. 34.

17.   Lamont, *The Philosophy of Humanism*, pp. 11-12.

18.   *Animal*. VIII: 1, 2.

19.   For a detailed analysis of the use of the term *methorios* in classical antiquity, see J. Daniélou, *L'être et le Temps chez Grégoire de Nysse*, pp. 116-132.

20.   Cyril of Alexandria, see *Patrologia Graeca* Vol. LXXV 853 C and Vol. LXXIII 1045 C.

21.   Ibid.

CHAPTER TEN

1.   Buddhist philosophy begins in India with the *Vaibhāsika* school with its realistic dualism. The *Sautrāntikas* developed different epistemological theories, more or less on a phenomenalist line. This was followed by the *Yogācāra* school which regarded reality as 'experience-ability'. *Mādhyamika* was the fourth school, which criticized all the three previous ones. The *Mādhyamikas* later divided into *Svātantrikas* and *Prāsangikas*.

2.   *Pratītyasamudpādahridayā*, See P. V. Bapat, Gen. Ed.: 2500 *years of Buddhism*. Government of India Publication, 1956, p. 425.

3.   The eight negatives are difficult to translate exactly. *Anirōdham, anutpadam, anunchhēdam, aśaśvatam, anēkārtham, anānartham, anāgamam, anirgamam*—Mādhyamika-karika.

4.   T. R. V. Murti, *The Central Philosophy of Buddhism* (London: George Allen & Unwin, Second edition 1960), p. 313.

5.   C. G. Jung, *Synchronicity. An Acausal Connecting Principle.* Eng. Tr. R. F. C. Hull (London: Routledge & Kegan Paul, 1972), p. 7.

6.   Op. cit, p. 49.

7.   Between 1955 and 1972, in fifteen years the Two-Third World share in world market economy exports fell from 28 to 19 percent and by 1977 it has fallen further to 17 percent. The trend is still downward.

8.   From $3600 million to $8500 million.

9.   V. I. Lenin, "On the Significance of Militant Materialism" *Collected Works* (Moscow: Vol. 33), p 233. For the copy theory, see his "The Theory of Knowledge" in *Collected Works*, Vol. 14 (1908), pp. 40–193. See esp. p. 105.

10.   *Philosophy in the USSR Problems of Dialectical Materialism.* (Moscow: Progress Publishers, 1977), p. 15.

11.   Hegel e.g. never said that only *our* mind really exists.

12.   Cited in Loren R. Graham, *Science and Philosophy in the Soviet Union.* (NY: Vintage Books, 1974), p. 40.

13.   Progress Publishers, Moscow, 1974.

14.   Op. cit. p. 204, italics present author's.

15.   Ibid.

16.   Op. cit., pp. 207–208.

17.   Op. cit., pp. 209–210.

18.   Lenin, "The Theory of Knowledge" in *Collected Works*, Volume 14 (1908), Moscow, 1972, p. 136.

19.   V. I. Lenin, "The Recent Revolution in Natural Science," in *Collected Works*, Vol. 14, p. 281.

20.   *The fundamentals of Marxist-Leninist Philosophy*, Moscow, 1974, p. 244.

21.   "Dialectics of Being and Consciousness," in *Philosophy in the USSR*, p. 43.

22.   Besides the well known $E = mc^2$ Einstein gave us the other formula for relating the mass of a body in motion to its mass at rest—viz.

$$\frac{m = m_0}{\sqrt{1 - \dfrac{v^2}{c^2}}}$$

where m is motion mass and mo is inert mass, V is velocity of body and c is speed of light in a vacuum.

23. Op. cit., p. 43.

24. Ibid., p. 51.

25. Ibid., p. 52.

26. Ibid., p. 58.

27. Melukhin, op. cit., p. 60.

28. Op. cit., p. 60.

29. The unity of all matter is an important axiom in Marxist ideology. So is the infinity and eternity of matter.

30. Op. cit., p. 63.

31. Op. cit., p. 66.

32. A magnitude arrived at by combining the gravitational constant G with Planck's constant and the velocity of light.

33. Diels, *Fragments*, Eng. Tr. in Kathleen Freeman, *Ancilla to the Pre-Socratic Philosophers*. OXford, 1971, p. 19

34. Ibid., p. 26.

35. In Christian usage the preposition "of" followed by a place-name means that the person named is the *bishop* of that locality. In the case of Śankara, Kāladi is simply his birthplace near Alwaye, in Kerala.

36. Gregory of Nyssa, *Answer to Eunomius' Second Book*, Jaeger ed. Vol. I, 275ff. PG 45, 969ff. Eng. Tr. NPNF Series Two, Vol. V, pp. 267ff.

37. Ibid., I:277, PG. 45:969c, NPNF V:268.

CHAPTER ELEVEN

1. Gateway Edition, Chicago, 1969. Maslow was Professor of Psychology at Brandeis, President of the American Psychological Association, and the founder of the Association for Humanist Psychology.

2. Op. cit. Foreword by Arthur Wirth, p. ix.

3. Op. cit., p. xiv.

4. Op. cit., p. 16.

5. For a slightly out of date but still highly useful journalistic Western study of this, see Ostrander, S., and L. Schroeder, *Psychic Discoveries Behind the Iron Curtain*, Englewood Cliffs, N.J., Prentice-Hall, 1970.

6. Vol VI, No. 2., p. 137 in article on "Man as the Unique Creator of Sense".

7. Janus Kuczynski, op. cit., p. 140.

8. Op. cit., p. 142.

9. See e.g. the American Marxist Howard Parsons' article on "Science and Technology: Means to What End?" in *Dialectics and Humanism* (Warsaw: Vol. VI, No. 2, Spring 1979), p. 73ff.

CHAPTER TWELVE

1. *La Philosophie Positive* par August Comte, résumé par Jules Rig, Paris 1880, p. 2. Eng Tr. *The Positive Philosophy of Auguste Comte*, Harriet Matineau, 2 Vols., London: 1853, Vol. I, pp. 1-2.

2. See op. cit., Eng. Tr., pp. 16-17.

3. In a letter to Father Mersenne in 1638, see R. M. Eaton, ed., *Descartes Selections*, New York:, Charles Scribner's Sons, 1927, cited by Wesley Salmon: "The Foundations of Scientific Inference in Robert G. Colodny, ed., *Mind and Cosmos*, University of Pittsburgh Press, 1966, p. 136.

4. F. Bacon, *Novum Organon*, aphorism XIX.

5. Op. cit., Eng. Tr., p. 33, capitals in original.

6. "Philosophy and the Scientific Image of Man in Robert G. Colodny, ed., *Frontiers of Science and Philosophy*, University of Pittsburgh Press, 1962, p. 37.

7. A question posed by Leibniz, repeated by Schelling, but made central by Heidegger.

8. *Was Heisst Denken*, 1954. Eng. Tr. *What is Thinking?* (New York, Evanston, London: Harper and Row, 1968), p. 18.

9. *Unterwegs Zur Sprache*, 1959.

10. In Michael Murray, ed., *Heidegger and Modern Philosophy*, New Haven: Yale University Press, 1978, pp. 35ff.

11. Heidegger has a chapter on "The Age of World-views in *Holzwege* (1950), Eng. Tr. by Marjorie Greene in *Boundary* 2(1976): 34-35.

12. See Ilya Prigogine, *Order out of Chaos, Man's New Dialogue with Nature*, (New York: Bantam Books, 1984) and *From Being to Becoming— Time and Complexity in the Physical Sciences*, (San Francisco: W. H. Freeman, 1980).

13. Rupert Sheldrake, *A New Science of Life: The Hypothesis of Formative Causation*, (Granada, London, New York, etc.: A Paladin Book, 1983).

14. See *Order out of Chaos*, pp. 286-287.

15. Oxford University Press, 1985.

16. Op. cit., pp. 80-81.

17. G.W.F. Hegel, "Glauben und Wissen in *Gesammelte Werke*, (Hamburg: Meiner, 1968), 4:315.

18. Chico, California: Scholars Press (Harvard Studies in World Religions No. 4), 1985. See especially the essay "Beyond Believing and Knowing, pp. 202-220.

19. Op. cit., pp. 204-205.

20. In "The Christian Attitude Toward Non-Christian Religions, *Zeitschrift fur Missionswissenschaft und Religionswissenschaft*, 51, 3(1957): 262, cited by Mehta, op. cit., pp. 210-211.

21. Mehta, op. cit., p. 215.

CHAPTER THIRTEEN

1. J.L. Mehta, *India and the West*, p. 183.

2. M. Heidegger, "Holderlins Erde und Himmel in *Erlanterungen Zu Holderlins Dichtung*, Frankfurt: Klostermann, vierte, erweiterte Auflage, 1971, p. 177. Eng. Tr. in Mehta, op. cit., p. 233.

3. Mehta, op. cit., p. 234.

4. J. Habermas, *Erkentris und Interesse*. Eng. Tr. *Knowledge and Human Interests*, (Boston: Beacon Press, 1971).

5. See his *Zur Rekonstruction des Historischen Materialismus*, Frankfurt: Suhrkamp, 1976. Four of the essays in this collection, translated by Fr. Thomas McCarthy, are available in English, in J. Habermas, *Communication and the Evolution of Society*, Boston.

GENERAL THEOLOGICAL SEMINARY
NEW YORK